To Papa,

from Chris
Feb 2000

PIAGET

Watches and Wonders
Since 1874

Franco Cologni　　Giampiero Negretti　　Franco Nencini

Piaget

Watches and Wonders

Since 1874

Abbeville Press Publishers　New York　London　Paris

Editor	Franco Cologni
Designer	Raymond Dandrieux
Graphics	Daniele Cipolat
Technical/historical consultant	Emil Keller
Editorial staff	Ezio Sinigaglia Grazia Valtorta Hazeline van Swaay-Hoog
Cover and jacket	Ateliers ABC
Acknowledgments to	Patrizia Baini Jean-Claude Barbezat Yvan Barbezat Yves Bornand Cédric Bossert Philippe Chalandon John Corlet Roger A. Dick Dominique Ferrero Francis Gouten Charles Guyot Samuel Hasler Barbara Soleyman Albert Kaufmann Danièle Marin Sébastien Page Sylvie Pernet Camille Pilet Pierre Rainero Richard Rod Editions Scriptar S.A. Time Products plc C.S. Torres S.A.

Piaget - Watches and Wonders
Franco Cologni, Giampiero Negretti, Franco Nencini

Copyright © 1994
by Piaget

First published in the United States of America
in 1995 by Abbeville Press, 488 Madison Avenue, New York, N.Y. 10022

All rights reserved under international copyright conventions. No part of this book may be reproduced or utilized in any form or by any means, electronic or mechanical, including photocopying, recording, or by any information storage and retrieval system, without permission in writing from the publisher. Inquiries should be addressed to Abbeville Publishing Group, 488 Madison Avenue, New York, N.Y. 10022.
Printed and bound in Italy. February 1996

ISBN 0-7892-0078-3

English translation: Lory Frankel

Publishing coordinator: Publiprom

Contents

1874 • 1945
Piaget before Piaget

In the village of La Côte-aux-Fées, the Piaget family begins making parts for watches, specializing in lever escapements. It is not long before Piaget movements are being used by Europe's most prestigious watchmakers. *page 9*

The Excellence of mechanical movements.

How a watch works. The special characteristics of Piaget's movements. Manual and automatic movements. *page 35*

1945 • 1960
The birth of a myth

Piaget decides to sign his watches and to build a new, advanced factory. In the mid-1950s, the ultra-thin 9P caliber appears, opening the range of models to new possibilities, especially those for ladies. *page 45*

Old or new: The quality remains constant.

From the tradition of skilled hands to the most sophisticated computer equipment: in La Côte-aux-Fées and in the Geneva workshop, the work of wedding tradition to modernity is carried on. *page 73*

1960 • 1970
The boom years

Piaget becomes an international trademark. A first showroom opens in Geneva. With the use of gemstones for the dials, the models grow even more exclusive and precious. During the same period, a line of jewelry-watches is introduced, facilitated by the integration of the Geneva workshops specializing in working with precious stones and making watch cases and bracelets. *page 83*

Time and the jewel

The painstaking work of great master watchmakers: embellishing a watch with precious gems and transforming it into a jewel-watch. Hundreds of hours of work, starting with a golden thread, to create a bracelet more pliable than silk. *page 121*

1970 • 1980

High Fashion, High Technology

The quartz revolution does not catch Piaget off balance, which develops and produces an ultra-thin electronic movement. Color dials, rich design, and quartz precision combine to keep Piaget at the top in the exclusive and highly competitive world of fine watches.   *page 131*

Quartz: Total precision

Quartz signals a revolution in the art of measuring time. The Piaget tradition endows the electronic with elements of refinement.   *page 171*

1980 • 1988

The luxury years

Quartz watches become ever more sophisticated; interest revives in the mechanical models that Piaget never stopped perfecting. The watch is a jewel: inside, by virtue of its technology; outside, by virtue of the originality and refinement of its conception. Piaget makes the most precious watches in the world.   *page 181*

Special models and one-of-a-kind Creations

From coin watches to special models made on order: a panorama of all the unique watches that delight collectors and enthusiasts the world over.   *page 219*

1988

A new era begins

The association of two illustrious names: Piaget and Cartier. New international development: a synergy conducive to the creativity of the two companies. This encounter marks the cultural changes of the 1990s and the introduction of Piaget's great collections for the year 2000: Tanagra, Polo, Gouverneur, which draw from the past while remaining undeniably contemporary.   *page 231*

This book is the history of a company, but it is a first and foremost the history of a family - my family.

It is the history of a company that began as a modest artisanal venture and moved steadily to the top of its field, so that its very name, Piaget, has become synonymous with excellence in watchmaking.

The expression of an art, born of one of the very few Swiss watchmaking companies over a century old, the name Piaget today symbolizes the idea of perfection in watchmaking, of exquisite craft in jewelry, and of inimitable design.

"Piaget Watches and Wonders" is equally the history of a family intimately connected with the development of the enterprise, from Georges Piaget, the founder, to myself, the fourth generation.

Each man who has succeeded to the leadership of the house of Piaget has made his contribution to its firm foundations, and without their unshakable faith, their unswerving dedication to the search for exellence, Piaget would not be the great company it is today, 120 years later.

The purpose of this book is to tell the story of this dynasty of watchmakers for the first time using rare historic evidence, including the recollections of the last living witnesses, and unpublished documents, including drawings and photographs from the company's archives, here offered to collectors, connoisseurs, and Piaget clients.

Piaget's Private Collections are indeed reproduced in this book.

The memory of the past serves to write the future ...

Yves G. Piaget

ing, demanded concentration and, like watchmaking, a great instinct for opportunity – smuggling contraband. This is how they earned the local nickname of the "Niquelets," from their ability to "faire la nique," that is, to "dupe the customs" by crossing the frontier secretly with entire herds of cattle, sheep, or, perhaps, fairies (given their invisibility) ...

Clockmaking thus made its appearance in the canton of Neuchâtel toward the end of the seventeenth century, where it found a fertile ground, all prepared and ready to thrive. A statistic of 1848 tells us that among the entire population of the canton, fully 23 percent actively worked in this sector, and the percentage exceeded 40 at La Chaux-de-Fonds and Le Locle. In the district of Val-de-Travers, the number that year (in the middle of the century) came to the average of the canton: 21 percent. The work was carried out strictly in artisanal fashion, farmed out to numerous workshops each one contributing to making watches in manufacturing certain detachable parts known as *parties brisées* (literally, broken pieces). An industrial census taken at the beginning of the 1870s found in La Chaux-de-Fonds alone at least forty different "specialist" professions dealing with the making of watches. The small workshops produce the raw parts for the larger factories, especially those located in Geneva, which polish, perfect, and assemble them.

However, the quality of the production taking place throughout the area was not always of the highest level. Many artisans working in small or tiny shops contrived to turn out by themselves finished watches using rudimentary tools, and the results were far from refined. On the other hand, simple as the mechanism was, the watch known as the Roskopf, mass-produced beginning in the late 1860s, worked well and was enormously successful (tens of millions sold throughout the world); it became the "popular" watch par excellence.

An unexpected shift occurred around the mid-1870s. First of all, in 1874 a revision of the Federal Constitution of 1848 finally brought Switzerland into the modern age, ready to make the leap into the industrial and commercial development that would be under way by the end of the century. While main-

The armorial bearing of the commune of La Côte-aux-Fées dispels any etymological doubts about the name's source: definitely Hill of the Sheep, not of the Fairies.

taining the administrative independence of the cantons, Switzerland became a federal state in the true sense of the word, capable of a unified approach to politics, economics, and industry and definitively free from the obstacle that the anachronistic customs collection between cantons had imposed. This would prove to have far-reaching implications for the future of clock- and watchmaking in Switzerland.

In the second place, a severe crisis suddenly affected that nascent industry. A single statistic, impressive in its seriousness, suffices to give an idea of the speed and proportions of the downfall: in 1872 Switzerland exported the trifling number of 366,000 watches to the United States; in 1876, barely four years later, only a fifth of that number, 70,000, went across the Atlantic. As might be expected, a quick remedy was sought.

In fact, the impressive growth of the watch industry in the United States had temporarily forced the Swiss watch industry to its knees. In 1876, a great international fair was held in Philadelphia, to which Swiss producers sent representatives - Jacques David of Saint-Imier and Edouard Favre-Perret of Le Locle. On his return, Favre-Perret called a conference at La Chaux-de-Fonds that left the participants literally stupefied: all the assumptions on which the myth of the superiority of the Swiss watchmaking industry had until then been built seemed to collapse in the face of his startling revelation: the Swiss watch of average quality was decidedly inferior in terms of precision to any American mass-produced model that sold for a few dollars.

American watchmaking was on its way to supplanting the European industry, particularly in Switzerland, by virtue of more modern and efficient working methods. Whereas production in Switzerland was scattered over a wide area and diffused into hundreds of small and tiny workshops, each one specializing in the making of one part, in America it was centrally organized. Large factories gathered in one place all the labor necessary for the different stages of production, which was heavily mechanized. This was the future of watchmaking, and Switzerland had to adapt to the profound transformation it implied.

Switzerland adapted only in part, and that was the secret of its recovered and increased fortune. The small revolution that took place at the end of the century quickly effected a useful differentiation of production based on the targeted market share. The production aimed at the middle market adopted the American model: it fit itself out with modern machinery and tools, centralized its organization as much as possible, and noticeably raised its products' level of quality - a paradox only in appearance. It came about because the workers, although now employing machines, had always been very specialized, and they were in a position to intervene during each phase of production to improve and perfect the product. However, the high end of production continued to take advantage of the artisanal work organization, which ensured an unequaled level of quality. Only the less qualified among the multitude of small workshops that up to this time had constituted the fabric of the Swiss watchmaking industry disappeared. The best among them not only survived but drew from these events a new strength: while the production of ordinary watches was concentrated in the large factories, the small artisanal workshops, turning their greater flexibility to their profit and refining their manual skills and know-how, from that moment became centers of technical and aesthetic innovation.

In this period of fermentation, appropriate to the onset of a crisis and the rebirth that follows, the Piaget

The most famous of the "Piagets before Piaget": Alexis-Marie Piaget (1802-1870), president of the provisional government of 1848 and then president of the State Council for twelve years.

company was created. The Piaget family originated in Val-de-Travers; the name crops up repeatedly in deeds and documents of the communes of Les Bayards, La Côte-aux-Fées, and Les Verrières during the fourteenth century. The old spelling of the name is Peaget or Péaget, which leads us to surmise that their first occupation was as toll collectors (*péager*).

There are, of course, numerous branches of the Piaget family, one of which settled at La Côte-aux-Fées in the sixteenth century. Other Piagets can be found in Geneva, in the canton of Vaud, in nearby Neuchâtel, and in the faraway United States of America. Alexis Marie Piaget (1802-1870) came from Neuchâtel. In his youth, he ran a printing office in Paris, then returned to his hometown, where he established a law office and forged a brilliant career – deputy to the Legislative Assembly, president of the provisional government in 1848, president of the State Council for twelve years, and finally as deputy in the house of representatives, one of the leading actors in the reorganization of the federal state. The most famous Piaget, Jean, the great psychologist of the formative period in human development, also came from Neuchâtel. He was a professor at the University of Geneva and the author of many seminal books, including *The Construction of Reality in the Child* (1926).

Among the predecessors to the house of Piaget, many were involved in watchmaking before the birth of the manufactory that would add such luster to the name. In reality, it was hard to live in the area of La Côte-aux-Fées and Les Verrières without having at least one watchmaker relative. The master watchmakers of the region were already famous for the quality of their work at the end of the eighteenth century, to the extent that Abraham Louis Breguet, the brilliant

Georges Edouard Piaget (1855-1931), the founder of the Piaget company.

watchmaker to Louis XVI and Napoleon, entrusted the making of certain pieces, which demanded fine precision and an extraordinary attention to the smallest details of manual work, to the artisans of Les Verrières.

Evidence that the Piaget family made watches before founding its company is definitively offered in the form of four pieces, today in the Piaget collection: four pocket watches signed Piaget, dating from 1820, 1825, 1830, and 1836 - that is, several decades before the founding of the Maison Piaget in 1874, and even decades before the birth of its founder, Georges (1855). The piece from 1820 is of particular interest. A pocket watch with repeater at the hour and quarters from La Côte-aux-Fées, it is clearly "related" to the celebrated ultra-thins that appeared a century and a half later.

In any case, Georges Edouard Piaget founded the company that bears his name and opened his first watchmaking workshop at La Côte-aux-Fées in 1874, so this is the official date of the birth of the Maison Piaget. In 1874 Georges was nineteen, a truly precocious age at which to launch a business. But Georges had a liking for adventure: he was a dreamer with a practical bent. As a child, he hankered after a life of travel and bold enterprises: the United States, perhaps Canada. The family's modest means convinced him to abandon his dreams: while still quite young he became apprenticed to a watchmaker of nearby Les Bayards. There he not only found himself learning the secrets of the trade but he was also able to experiment with an innovative technique, which he turned to profit when he opened his own workshop at La Côte-aux-Fées.

The watchmaker to whom Georges was apprenticed specialized in the production of a type of escapement that, in point of fact, was not strictly speaking a novelty, as it had been invented a century earlier. However, the lever escapement had been "rediscovered" only at this time. It would prove to be a great success. The "planters of escapements," that is, the artisans who make escapements and place them in the mechanisms, have been (and, no doubt, will for a long time continue to be) among the rare "master builders in small" of watchmaking without whom the great companies could not function, even after the significant shift to mechanized labor. The "little secret" of Georges Piaget thus constituted a perfect basis for the founding of a business.

In the mechanism of a timepiece, the escapement plays a fundamental role: it is the device that regulates the flow of power generated by the mainspring and, at the same time, furnishes the regulating mechanism, or balance, with energy given out in oscillations, which keeps it in movement. For centuries, only one kind of escapement, the verge escapement, was used, beginning in the thirteenth century, for the first clocks in towers and steeples. Starting in the eighteenth century, with the spread of personal watches (pocket watches for men; brooches or châtelaines for women) and wall or standing clocks, many different types of escapement systems were introduced. Two of these proved indispensable to the personal watch (a term that encompasses pocket and wristwatches): the cylinder escapement and the lever escapement.

The cylinder escapement was invented around 1725 by the Englishman George Graham and was widely used until the end of the last century. It owes its name to its form: a notched cylinder that constitutes the balance staff with cavities that receive the triangular teeth of the escape wheel. Technically, it is what is called a deadbeat escapement, allowing little or no recoil. The lever escapement, on the other hand, is called "free," or detached, as it gives the balance almost complete freedom to oscillate.

Invented by the Englishman Thómas Mudge in 1754, the lever escapement had been forgotten for almost a century. It was not until the 1830s and 1840s that it was rediscovered by certain watchmakers of Geneva, especially Georges Leschot and his colleague with a remarkably similar surname, Antoine Léchaud (1812-1875), who improved it and gradually made it reliable. Because of its technical characteristics, the lever escapement contributed decisively to the development of precision timekeeping in Switzerland.

Not satisfied with simply supplying the escapement to his commissioners, Georges Piaget and his work-

A pendant watch that belonged to the wife of Georges Piaget, founder of the company. For the first time, the inscription "Côtes-aux-Fées" appears, paired with the signature of the company. Case in 18-karat gold, cylinder escapement. Made 1850-60.

shop made parts of the movement and, on request, carried out an entire series of operations that transformed the half-finished piece, come from the maker or made at La Côte-aux-Fées, into a finished watch. Little by little, Georges specialized in the making of entire movements as well as finished watches that he would provide to the manufacturer, who had only to add his trademark to complete it. For several decades, the Maison Piaget thus carried on a work both highly skilled and very modest: it made movements and watches but did not sign them. This makes it more difficult to reconstruct a detailed narrative of the company in the early period of its existence.

Nevertheless, it is not difficult to picture the way in which the Piaget family worked. There was no clear delineation between the house and the workshop. A single structure contained living space on the lower floors and the workshop mostly in the attic, because of the greater amount of light it afforded. A workday might be as long as fifteen hours. All the members of the family were involved in some artisanal activity, from the older members, who often had the most useful knowledge of how to do things, to the children, who quickly learned how to take advantage of their small hands in precision operations calling for the utmost delicacy. Even the women worked, during the hours when they were not carrying out their domestic duties. If they sat down to knit or darn, they engaged the other members of the family in conversation, probably concerning the life of the village. After all, the large windows let in not only light but also the noise of wagons, and a continuous stream of news, from the smallest item to the gravest matter, was silently passed on from neighbor to neighbor as each noted the comings and goings in this tiny mountain village.

The Piaget family surely found itself the subject of scrutiny as well, especially as it gradually increased in size. In 1881 Georges married Emma Bünzli, and several children soon followed: in 1882 Edouard, in 1883 John, in 1884 Marguerite, in 1885 Timothée, then Lydie, Anna, Willy ... for a total of fourteen. If succession to the family business seemed to be guaranteed, there remained the problem of finding room for all.

Although it was not put together until 1825, this is probably the oldest pocket watch signed Piaget, as indicated by the movement, still a conoid type, with verge escapement. The watch has a case in 18-karat gold and a repeater that strikes the quarters (which works by pressing the small button on the side of the case and pressing on the pendant) and is wound and set by means of a key. Enamel dial.

Edouard Kaiser (1855-1931), The Watchmaker (1898).

By a stroke of luck, the Evangelical Free Church, which the Piagets attended faithfully, built a large chapel in La Côte-aux-Fées in 1890. The watchmaking workshop was able to transfer its operations to the ground floor of that structure, in one of its largest and most comfortable rooms. The document bearing Georges Piaget's signature used for the cover of this book dates from this period (21 January 1890); it is a subscription of one hundred Swiss francs for the benefice of the Evangelical Free Church. It may have been in homage to the German origin of the Protestant Reformation initiated by Luther and brought to Switzerland by Huldrych Zwingli that the founder of the Maison Piaget chose to write his first name in its German form (Georg) rather than its usual French form (Georges).

By 1911, the spaces offered by the Evangelical Church had grown insufficient. With the purchase of the Café Français, the company gained plenty of room to accommodate its increased business, and scope for its expansion. Yet Piaget still clung to its "noble anonymity," which seemed to suit perfectly the dynamic and efficient but also extremely serious and discreet nature of its founder.

Georges Piaget died in 1931, at the age of seventy-six. During his last ten years he became blind as a result of diabetes, although it did not prevent him from taking a lively interest in his company's affairs. His fourth child, Timothée, succeeded him at the helm in 1911. Soon after, the small family business was transformed into a partnership, which included Georges himself, his brother William, and three of his sons - Timothée and his two older brothers, Edouard and John.

Georges thus was able to witness and collaborate in the first major expansion of the business, which remained unknown to the public for many years yet but, on the other hand, had made a name for itself among the watchmakers of Geneva. This is borne out by the account books, which toward the end of the 1920s listed the most famous names in Swiss watchmaking among its clients: from Rolex to Breguet, Omega, Zenith, Vulcain, Longines, Niton, Ulysse Nardin, Ebel, Vacheron & Constantin, and Audemars Piguet. Outside of Switzerland, it did business with the legendary firm of Cartier.

From its peaceful hermitage of La Côte-aux-Fées, Piaget supplied the great Geneva houses with parts, complete movements, and sometimes even finished watches, but it signed none of its products.

That division of labor between city and country comes from a tradition dating back to the end of the seventeenth century. La Fabrique – the quasi-mythical name that designates all of the arts and crafts that go into clock- and watchmaking in Geneva – had always been solidly dominated by the upper middle class of Geneva, which left the production of parts and unfinished movements to the growing network of very small businesses in the nearby countryside and concentrated instead on the work of finishing and assembling the watch. During the eighteenth century and into a good part of the nineteenth, this state of affairs was fixed in place by a series of protectionist laws and regulations blocking anyone from entering the "noble ranks" of La Fabrique who did not belong to the fine flower of the Geneva bourgeoisie. Although changes occurred gradually over the course of many decades, the effects of this centuries-old division of labor were still noticeable at the beginning of the twentieth century.

Meanwhile, Geneva turned into an international city. A certain cosmopolitanism had always been part of its character, since the beginning of its success as the watchmaking capital dated from early in the six-

This much more modern watch dates from 1820. The movement, already a Lepine caliber, employs a lateral lever escapement, and one of the first antishock devices is found on its balance. Repeater at the quarters and key winder. Silver dial, engine-turned in its center, with Roman numerals engraved on the outside, and 18-karat-gold case.

teenth century, when the Calvinist city freely welcomed the French Protestants persecuted in their country by the Catholic king. The very close ties between Geneva and France remained strong and fruitful over the centuries, opening the city on the Lake Leman (or Lake of Geneva) to European vistas. Geneva also lent its most famous son, Jean-Jacques Rousseau, to the glorious period of the French Enlightenment.

In the second half of the nineteenth century, Geneva's international vocation gained one of its most significant consecrations with the signature, in 1864, of the first Geneva Convention on the treatment of prisoners of war, the protection of wounded soldiers, and civilian victims of combat. The agreement between the major powers was arrived at through the active interest of the Red Cross. It represents one of the fundamental elements of contemporary society as humanitarian progress.

Thirty years later another great event, the National Exposition of 1896, would give the city a new face, more modern and dynamic, which would serve as a good preparation for the twentieth century. Around the alleys of the old city and the Faubourg Saint-Gervais, the veritable home of La Fabrique, rose up new buildings of the bourgeois prosperity on wide avenues and along the lakefront, with its magnificent views.

Especially after World War I, Geneva would see its role as the "supranational capital" elevated with the creation of the League of Nations, which met for the first time on 15 November 1920 on the lakeshore. With the arrival of representatives from countries all over the world and the flowering of international

This model with a silver case dates from around 1825. "H. Piaget – Rieff à Yverdon" appears on the dial. Verge escapement and winding and setting by means of a key. "Embossed" dial corresponding to the numerals permits one to read the hour through touch.

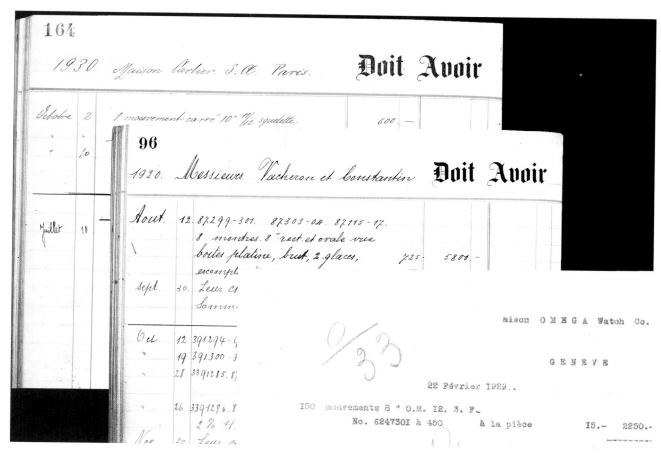

diplomacy, a period of new splendor began, for La Fabrique as well as for the entire city.

On the majestic and twinkling quays of the lakefront, large hotels opened their discreetly silent doors and their richly carpeted salons. An intense and elegant social life revolving around international professionals on a high level, called to Geneva to discharge their diplomatic and ministerial responsabilities, began. "Foreign affairs," so to speak, provided an extraordinary impetus to "internal affairs." Watches and jewelry, longtime products of the city, were also most likely to attract new clients, rich aristocrats, and wealthy society ladies.

Geneva turned itself into a sparkling city of shop windows. One after another glittered with gold, precious stones, diamonds. Georges Piaget, before his illness caused him to lose his sight, surely would have taken the time to be seduced by the transformation

The rare bookkeeping documents conserved in the Piaget archives confirm that during the 1920s and 1930s, the small factory at La Côte-aux-Fées counted among its clients the elite of the Swiss, as well as foreign, watchmakers. The page from the account book of 1920 posts the account of a client with a famous name – Vacheron & Constantin, for whom, we learn, Piaget made complete movements and even (at the top of the list) finished watches. The page from the 1930 account book shows an exceptional client, which has become of great importance to Piaget's recent history – Cartier, for which Piaget supplied a rectangular skeleton movement on 2 October 1930. The third document, dated 1929, comes from an account book that placed together copies of bills. This one is addressed to the Omega Watch Co. for furnishing it with no fewer than 150 complete movements.

of Geneva's streets into fabulous roadways of the Thousand and One Nights. His sons certainly would have. All of the most important clients of the Piagets were located in Geneva, and under the pressure of this new golden age (of watchmaking as well), they multiplied their orders. This would have meant making the trip to their "capital" more often, which had become much easier with the advent of the automobile. It was perhaps in front of those shop windows, where watches shone as jewels, dressed up in diamonds, presenting their bracelets in extraordinarily flexible "golden links," that the idea took hold of Georges Piaget, or one of his sons, of the ambitious project that before long would transform the obscure company of La Côte-aux-Fées into the most famous watchmaker-jeweler in the world.

It must be that the anonymity of La Côte-aux-Fées began to seem too restricting to the young Piagets, while the attraction of Geneva exerted a strong force. Still, they waited another fifteen years after the death of the company's founder to decide to make the great leap; papa Piaget had taught them never to overestimate their strength.

Edouard Kaiser (1855-1931), Case-making Workshop *(1898). In many of his canvases, this Swiss painter has chosen to depict realistically artisans at work, illustrating in minute detail the different skills that go into watchmaking.*

The Pont du Mont-Blanc and the Quai des Bergues in Geneva as they appeared on 15 November 1920, when a crowd filled the streets to watch diplomats and politicians from the world over arrive for the first meeting of the Assembly of the League of Nations.

Left: Movement of the watch shown on page 18. Note the "signature" and the series number.

This aerial photograph of Geneva was taken in 1896, undoubtedly from a hotair balloon anchored to the ground, which towered over the National Exposition. The growth of the city is indicated by the clear contrast between the old, crowded-together roofs of the small houses in the old part of the city (in the middle of the photograph, near the lake shore) and the wide avenues and large apartment buildings constructed around it.

Signed Vy Piaget, London, this model from 1836 (made for the English market) features a finely chased gold case (restored in 1970) and a seconds counter. The movement, which is wound and set with a key, employs a duplex escapement.

Technical drawing showing the orthogonal projection and elevation of the lever escapement that Piaget was still using in 1922. Essential to the development of first the pocket watch and then the wristwatch, this type of escapement provides a high degree of precision without resorting to more complex, delicate, and costly devices. Piaget was among the first Swiss watchmaking companies to favor this device, which had been invented by the Englishman Thomas Mudge.

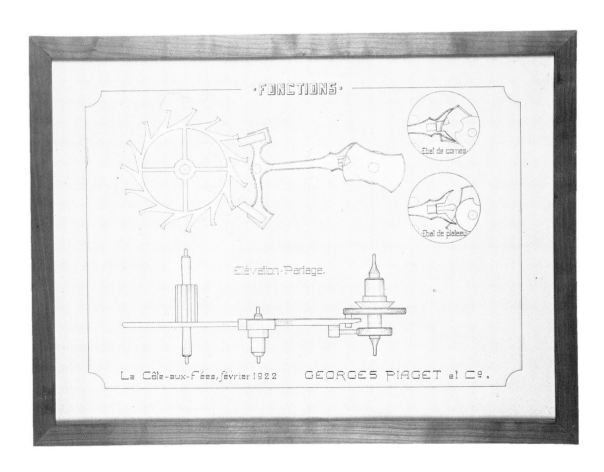

1

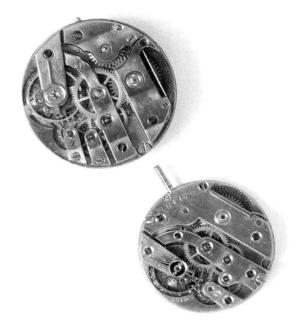

2

3

A group of several movements that were finished and assembled at La Côte-aux-Fées between 1914 and 1926. From above left: 1) an extra-thin movement (2.4 millimeters) for pocket watches; 2) and 3) two movements of reduced dimensions intended for ladies' watches; 4) a rectangular movement for a watch of that shape made in 1926; 5) one of the movements "en blanc," that is to say, partially mounted but still unfinished, that in 1920 Piaget received from other houses to finish and assemble the escapement and mount the rubies; and 6) the result of tens of hours of work and hundreds of operations.

4

6

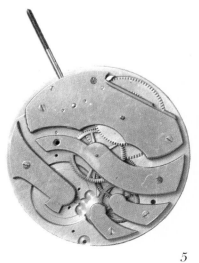

5

Made between 1830 and 1835, this example is signed "Piaget à Nismes" on the dial. It has a silver case and still uses a verge escapement. Winding and setting with a key, enamel dial.

The Excellence of Mechanical Movements

The mechanical watch is not a modern invention but an idea that goes back seven centuries. Surprisingly, it has changed little since then, as it continues to function according to the same principle. However, this is true only in theory; in actuality, the watch we wear on the wrist today is the result of a series of intuitions, innovations, and inventions ranging from mechanics to physics, from metallurgy to chemistry. Like the clocks of the thirteenth century, watches are composed of a "motor," a wheel train, an escapement system, and a balance. The "motor," or motive force, originally consisted of nothing but gravity: weights attached to ropes supplied enough energy as they descended to drive the wheel train. This then transmitted the motive force to the successive wheels that moved the hands, by distributing and gearing down the initial force. The escapement, by allowing only a small part of energy to "escape" at a time, transformed the continuous movement of time into intermittent movement. The balance, which is the oscillating section of the movement, received an impulse that

The 9P caliber, engineered in 1956 and manufactured the following year. With a diameter of 20.5 mm and a thickness of 2 mm, this movement remains one of the most refined and reliable "time machines" on the highest level of fine watchmaking. With 18 rubies, the 9P caliber has a balance that vibrates 19,800 times an hour.

fed the regulating movement. Today's mechanical watch works in the following manner: instead of weights, there is a spring; the power unit no longer weighs several hundred pounds but only a fraction of an ounce; the escapement and the regulator, once made of wrought iron, are now masterpieces of mechanical precision. And despite its ever-decreasing dimensions, the watch has become even more precise. In 1300, the large clocks placed on towers and steeples could be off by as much as twenty or thirty minutes a day. After the spiral balance spring, or hairspring, and the lever escapement came into use, a good pocket watch, by the end of the nineteenth century, lost no more than twelve seconds in a twenty-four-hour period. It was, in fact, the fabrication of the lever escapement that brought Piaget to fine watchmaking. The making of this par-

The famous automatic 12P caliber seen from above and in profile. With a diameter of 28.1 mm, the movement is 2.3 mm "high" - a record that remained unsurpassed for many years. Note the small oscillating weight (in 24-karat gold), supported by an inward curving bridge and placed on the periphery of the movement to reduce the thickness as much as possible. As can be verified, the rotor projects slightly outside the movement.

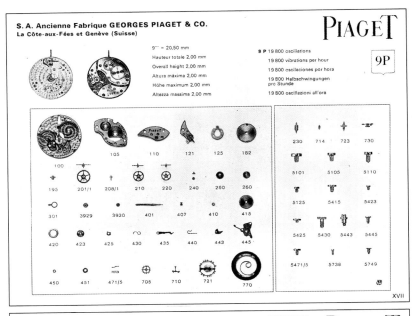

Tables showing the parts for the 9P and 12P calibers. It acts as a kind of spare parts catalogue, sent to all dealers and used by watch repairers to order the proper part from the Piaget company in order to ensure that the movement's correct functioning is not impaired and that it works exactly as it did originally. Thousands of spare parts, including some belonging to the oldest movements, can be found arranged in the drawers of Piaget's "Supplies" Department.

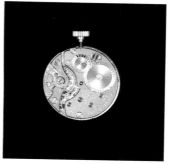

Another famous movement, with manual winding: the 4P caliber. Born in April 1976, this movement, of greatly reduced diameter and thickness (respectively, 14.2 and 2 mm), was primarily used for ladies' models. Aside from its high-frequency balance (21,600 vibrations), this caliber is distinguished by the contoured bridge of the balance and by the large plate that supports all the other components.

ticular type of escapement demands a series of complex operations culminating in the insertion of two small rubies shaped as trapezoids, which must be set with great precision and polished. To the specialized production of lever escapements, the Piaget company soon added that of other parts of the watch – parts that, as we have seen, were supplied to many other companies – to the point where, from the early years of the twentieth century, it made complete movements for wristwatches. Two elements characterize a "caliber" (the technical term used to designate a watch movement): its shape and dimensions. The latter – indicated by "lignes," an ancient French unit of measurement equivalent to 2.255 millimeters – has continuously grown smaller and provides a good example, better than any other element, of the technical level attained by the master watchmakers of La Côte-aux-Fées: in the smallest models, intended for ladies' watches, dozens and dozens of components work in synchrony in a space that barely exceeds a cubic centimeter. The 2P caliber, for example, measures 18.6 by 8.3 millimeters, with a thickness of 3.6 millimeters. Its tiny balance, that is, the oscillating unit that regulates the watch's advance, vibrates 21,600 times in an hour. The baguette-shaped 2P caliber offers another refined trait: the winding mechanism, which allows the watch to be wound and set, is placed not on the side but under the movement itself. With this movement, as well as with the caliber 6N, which has a round shape but also has its winding mechanism on the back, Piaget created several elegant models whose shape is not interrupted by the usual position of the winding crown. Among the best-known movements that Piaget conceived and fabricated are the calibers 4P, 5P, and 5N, respectively, round, rectangular, and barrel-shaped, as well as the calibers 6 lignes, round, 6P, and 6N (the latter has its winding mechanism on the back). However, the two most frequently illustrated "time machines" are the calibers 9P and 12P, created in the 1950s, which today continue to drive many Piaget models, although in modified form. With the 9P caliber of 1956, Piaget opened the door to a new generation of watches: the ultra-thins. Barely 2 millimeters thick, the 9P caliber, with a round shape, measures 9 lignes in diameter, that is, 20.5 millimeters. These reduced dimensions enabled it to be used for ladies' watches as well, while its precision, reliability, and strength have earned it a place in the pantheon of fine watchmaking. The 9P caliber had hardly begun its extraordinary career when Piaget put in the works another novelty, this one revolutionary. The request for a patent on a watch with an automatic winding device carries the date 8 May 1958. These innovations, carried out under Valentin Piaget, were aimed at the further slimming of the movement. Thus, the rotor of the winding mechanism was no longer located above the mechanism but "integrated," and the bridge that supports its arbor was placed on the same level as the other bridges of the movement. The rotor, made in 24-karat gold to add to its mass, is 1 millimeter thick and turns partly in the interior of a kind of niche made in the side of the case. Without going into greater technical detail, it can be said that Piaget's researches were realized in the form of an automatic movement of incredibly

The fascinating 2P caliber, also called the "baguette" movement after its shape. It is 18.6 mm long, 8.3 mm wide, and 3.6 mm thick. Aside from its special shape, its movement features a winding crown (which also serves to set the time) placed under the movement rather than on its side. These watches thus have an extraordinary elegance, as the shapes of the cases are not interrupted by any exterior element.

reduced thickness: only 2.3 millimeters. The patent was registered on 30 June 1959, and the following year the first watches driven by the 12P caliber went on the market. The public, the experts, and the press all pronounced it close to a miracle; no one had dreamed that it was possible to make a selfwinding movement of such slimness and offering a power-reserve of 36 hours.

As in the case of 9P, the balance of the 12P caliber vibrates 19,800 times an hour, which means that it is not a high-frequency movement. Nonetheless, because of the quality of its components, the precision of the assembly, and the care taken with its regulation, the caliber guarantees a precision comparable to that furnished by a chronometer and is less susceptible to wear than movements of higher frequency. Several variations on the 12P caliber followed the initial version, including the 12PC, which also featured date display.

CONFÉDÉRATION SUISSE

PROTECTION DES INVENTIONS

BREVET PRINCIPAL N° 339571

Le bureau soussigné, ayant constaté que les conditions requises par la loi sont remplies,
a délivré le présent brevet principal à la

Complications S.A.,
La Côte-aux-Fées (Neuchâtel),

pour l'invention décrite dans l'exposé ci-joint.

Date du dépôt de la demande : 8 mai 1958, 17 1/2 h.

La durée de la protection légale s'étend au plus jusqu'à l'expiration de 18 ans, à compter de la date de dépôt de la demande.
Les annuités échoient au jour du dépôt de la demande.
Le brevet est délivré sans garantie de l'Etat.

BUREAU FÉDÉRAL
DE LA PROPRIÉTÉ INTELLECTUELLE
LE DIRECTEUR :

BERNE, le 31 août 1959.

The first page of the patent issued for the 12P caliber by the Federal Bureau of Intellectual Property of the Swiss Confederation. After little more than a year of studies and comparative trials, the federal bureau officially recognized the technical importance of the caliber's automatic winding system. The originality of the peripheral rotor was particularly noted: the first major step toward reducing the thickness of the movement, it was also adapted to the extra-thin watches produced by Piaget.

1945 · 1960

A very elegant extra-thin pocket watch, distinguished by its unusual triangular gold case and by its back in cloisonné enamel decorated with multicolored floral motifs. The movement is the 9P caliber (1960).

Piaget: The Birth of a Myth

Post tenebras lux: After the shadows, light. This inscription, printed on the cover of the catalogue to the exhibition "Watches and Jewels" held in Geneva in 1944, was extremely emblematic of the difficult period experienced in neutral Switzerland during the tragic years of World War II. But the accent was placed on the light, on the desire to recover, and, in watchmaking, on the return to splendor and elegance of forms, materials, and technological progress. The years from 1940 through 1945 are often synonymous with a severe, square kind of watchmaking, in an almost neo-Gothic style; the greater part of the period's production, including such famous and worldly models as Reverso and the small Rolex Oyster, were made in "staybrite" steel. Even the two-thousandth anniversary of Geneva in 1942, noted the *Journal Suisse d'Horlogerie et de Bijouterie,* "witnessed a Geneva that suffered from claustrophobia. Surrounded by a continent darkened by war, the city of Calvin vegetates and runs the risk of forgetting the prodigious wealth of its heritage in the realm of clock - and watchmaking."

In reality, from the perspective of the rest of the world rather than that of the lake shore, things were not so bad. "The interruption caused by the war," wrote Harvard historian David S. Landes in *Revolution in Time*, "simply enhanced the Swiss domination of world markets. Everywhere else production of civilian timepieces more or less stopped while manufacturers devoted all their resources to producing war material... When the war ended, military orders for watches dropped abruptly, but unlike the years after the First World War, there was no depression, and long pent-up civilian demand pressed hard on Swiss watchmaking capacity. In the late 1940s merchants were ready to pay a premium for

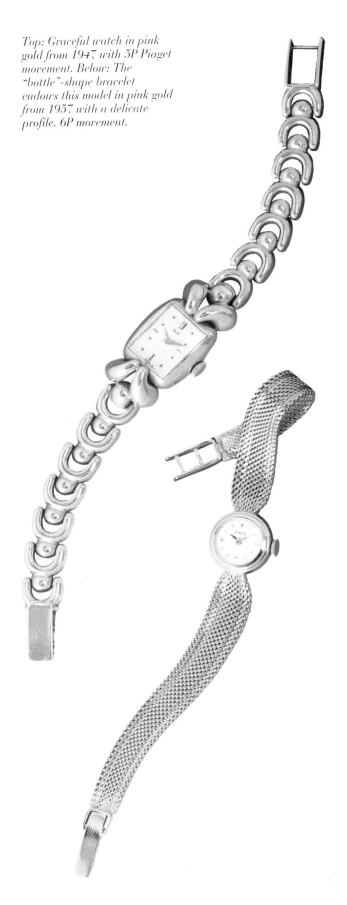

Top: Graceful watch in pink gold from 1947 with 9P Piaget movement. Below: The "bottle"-shape bracelet endows this model in pink gold from 1957 with a delicate profile. 6P movement.

Swiss watches... For a brief moment, the Swiss share of the world production is estimated to have surpassed 80 percent. It fell back rapidly as the crippled national industries of Japan, Germany, France, and the Soviet Union returned to production. Even so, the early fifties still found the Swiss with over half the world market."

The young Piagets, who had for many years lived under the same cloud as Geneva (since that was where their highest-level clients were, and, as we have seen, Geneva was more than ever the world capital of watchmaking), in the mid-1940s made the two basic decisions that would propel the Maison Piaget toward a great future and goals unthinkable to their ancestors. The main actors in this turning point were Gérald and Valentin, two of the twelve children of Timothée Piaget, who had diligently prepared, and in complementary ways, as it turned out, to succeed their father and their grandfather Georges.

Gérald had not only a fine intuitive sense but also a solid background in business and financial matters. From an early age, he was used to traveling to sell the products of the family company (at the same time, he turned his mind to researching people and places most appropriate to create, in a short time, a prestigious distribution network). In 1944 he was named general manager, and he naturally continued to benefit from the experience of Timothée, as well as John and Edouard, who formed the Board of Directors. That same year, Camille Pilet, an exceptionally dynamic man, joined the company. As Gérald's right-hand man on the business side he became a great traveler and made a decisive contribution to the development and success of the product in all the markets.

Valentin was an extraordinary technician, in which he was closer to his father and the world of the factory at La Côte-aux-Fées. He earned a lasting place in

World War II ended and humanity could once again dream of beautiful things. Gérald Piaget, at the end of the 1940s, underwent his umpteenth airplane flight to acquaint the world with the creations of Piaget.

the history of watchmaking for his invention and realization on an industrial scale of the famous ultra-thin 9P movement, which appeared in 1956 and constituted, as we will see, a true revolution.

Looking back, the decisions made by the Maison Piaget between 1940 and 1945 seem to be the fruit of a precise and clear-sighted strategy. The first, implemented from 1940, was to sign its own watches, no longer limiting itself to making movements for other companies, no matter how distinguished. To make the new name and its products known while awaiting the publication of a catalogue, advertisements were placed in professional journals. For example, a full-page advertisement appeared in February 1942 in the *Journal Suisse d'Horlogerie et de Bijouterie*, showing four glamorous models in gold from the coin series and carrying the message, "Luxury and Precision - Piaget et Co, La Côte-aux-Fées. Specialties: watches, calendar display, water-resistance, automatic, coins, jewelry, ultra-thin." This message carries two particularly interesting points. The first is the descriptive label "luxury and precision," which indicates that Piaget had decided to place its signed watches at the high end of the market. The second is the mention of watch-jewels and ultra-thins, which at the time were equipped with ETA or other mecanisms (assembled and finished at La Côte-aux-Fées), clearly revealing the existance of a plan for the future development.

The event discussed in all the newspapers and which

represented a turning point, as much for Piaget as for the quiet area of La Côte-aux-Fées, was the opening, in 1945, of a new facility with the aim of setting up an advanced watchmaking factory, with research laboratories, technical services, and other refinements that the past master watchmakers of the company could not even dream of. The director of the establishment – listed in the register of businesses under the name "S.A. Ancienne Fabrique Georges Piaget & Cie" – was at first Timothée Piaget, assisted by his sons Gérald and Valentin. The new complex could give work to over two hundred skilled artisans, which would take care of the entire village and the best artisans in the region. Even after the arrival of new machines, it was the priceless touch of skilled hands that gave Piaget watches their unique qualities. In respect to the new plant, which would lead to the creation of sensational calibers, ultra-thin and automatic, inventions that would not enter the world until the second half of the 1950s, it is tempting to ascribe to Valentin Piaget an excess of clairvoyance. Actually, the decision to open a new facility responded to the ever more pressing commissions of the great clients of the period (with Omega, Certina, and Wittnauer at the head of the list), who required ever greater quantities of movements as the end of the war approached, to be ready for the inevitable boom that would follow. That same year, 1945, saw the passing of the baton to the third generation: Gérald became the designated president and director of the company, with Valentin at his side as vice-president.

The first two catalogues offered give us a glimpse into the beginnings of Piaget. The first, covering the years 1946-47, shows thirty-three watches. Next to glamorous "coins" (whose price ran from 1,000 to 3,000 Swiss francs, according to the historic and numismatic value of the original coin: 100 Italian lire, 20 American dollars, 100 Austrian kronen, 50 Mexican pesos, 10 British guineas from the reign of Queen Victoria, and so on) we find ultra-thin skeleton models for men, round watches whose form fulfills the purest 1940s style, and tiny watches for ladies, also inspired by the dominant design, some with beautiful classic bracelets in gold. If one word only could be used to sum up the impression of this first catalogue, one would speak of "class." The 1948 catalogue, much richer with about a hundred offerings (ranging in price from 51 to 3,000 Swiss francs, the majority falling in the 200-300 bracket, which today would be characterized as high), also represents a leap in the aesthetic quality of the pieces. The round models present a perfect purity of line and a style unquestionably Piaget (for example, the number 8177 of 1949, with an automatic Arnold Schild movement, or 11695-2, automatic, 18-karat-gold case, with calendar, water-resistant, antimagnetic).

The same could be said of the watches for ladies, which, besides a more modern production, showed six or seven examples that could be defined as "the mark of the jeweler," with delicate, highly imaginative bracelets, some with small diamonds inlaid to exquisite effect (for example, the model in pink gold of 1945).

After this series of important novelties, at the beginning of the 1950s the company applied itself to reinforcing its presence on the watchmaking market, which was undergoing a complete renaissance. The demand for fine Swiss watches in the United States, Germany, Italy, and other countries was such that what could be called a real "black market" opened up for the best models, as dealers and merchants proved willing to pay a premium or an extra charge to be assured of receiving them. Luxury automobiles benefited from the same phenomenon, another sign of people's desire to own objects that would revive a sense of elegance and joy after the horrors of the war and the privations of the early postwar period. Piaget, with its at the beginning small and modest display booth, participated in the Basel Fair as from 1946. At the 1952 edition of the fair a truly splendid model in the middle of the showcase caught everyone's eyes: in solid gold, its cover carried a heraldic symbol – an eagle with wings spread – and bars of rubies and diamonds. This piece, a unique example commissioned by a noble family, constituted another mark of the jeweler Piaget. (Offered at an auction organized by Antiquorum Geneva on 25 April 1993, it fetched 33,800 Swiss francs.) It was probably around this time that Valentin Piaget was working on his project of the ultra-thin 9 lignes movement, which would be finished and patented in 1956 and introduced in the 1957 catalogue.

T hat invention was the first decisive event that would bring Piaget from a limited circle of connoisseurs to public notice, starting with the inner circle of Swiss watchmaking, as discreet as it is jealous. The effects of the new caliber on sales was immediately felt, especially among the models for ladies: Piaget, having a very slim mechanism to work with – and with a diameter of 20.5 millimeters – decided to

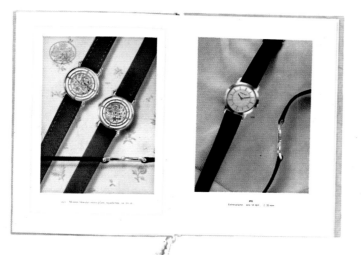

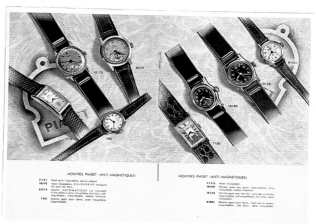

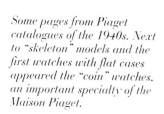

Some pages from Piaget catalogues of the 1940s. Next to "skeleton" models and the first watches with flat cases appeared the "coin" watches, an important specialty of the Maison Piaget.

Another very elegant pocket watch, distinguished by its triangular pendant and engine-turned back. The gold hour markings stand out in relief on the dial (1948).

widen the dials of ladies' watches. This had both aesthetic (it changed the design and the fashion) and practical repercussions, as older women could now read the time with perfect ease without putting on their glasses. Emil Keller, another valuable recruit of the Maison Piaget who, with Camille Pilet, worked in the upper echelons during the golden years, recalls his experience of this innovation as director of sales: "I remember perfectly having met at our store on the rue du Rhône many elegant ladies of Geneva, as well as American, French, and Italian women, who came to congratulate us on this invention, which finally did away with the embarrassing hunt for glasses in their handbags, then the annoyance of having to put them on and strain to read the time on the tiny little dials that were the rule in the forties and a good part of the fifties. With a Piaget on their wrists, they told me, looking at their watches gave them the pleasure of a simple, elegant gesture, accomplished with the nonchalance of ignoring their age…"

Presented at the Basel Fair of 1957, Piaget's 9P caliber created a sensation and struck envy among its competitors. "Since they could not criticize the technology or the value of the new movement," Emil Keller recalled, "they criticized our prices, saying that for us to ask the same prices as the great companies would be tantamount to cutting our throats." History soon proved that prediction wrong. Although few models with the new 9P caliber had been finished in time to be shown in the catalogue, there were already models worthy of admiration for their originality that conceded nothing in terms of elegance or refinement to those of Patek or Vacheron of the time. For men, there was the very beautiful square model in white gold with shantung engraved dial; for women, the large, round model with white gold case and black dial, both endowed with the new ultra-thin movement. It was not by chance that in this catalogue of 1957, so rich in innovations, next to the logo "Luxe et Précision" a second significant message read, "Piaget, the watch of the international elite." Not yet true, it would very quickly be accurate. The opening of Piaget's showroom at Geneva's most exclusive address, 40, rue du Rhône, proved crucial to making Piaget known to and prized by a new international elite (from the Beverly Hills society lady to the famous actress, from the commercial banker to the high-ranking United Nations official), which stayed at the Hôtel des Bergues or Richemond or Beau Rivage and, when in Geneva, never passed up the opportunity to visit the headquarters of the great watchmaking firms.

The new showroom opened a window on the world and, at the same time, served as a window through which the most important importers and clients could see and appreciate Piaget's watches, as they already esteemed the watches in the other exclusive showrooms nearby, such as those of Patek Philippe, Rolex, Vacheron & Constantin. Before the opening on 13 June 1959, the family and the firm's principal executives like Camille Pilet and Emil Keller, held intense discussions about the ideal qualities for the place to represent the company. It was, finally, Valentin Piaget's opinion that prevailed: it followed the philosophy practiced in the making of watches: be innovative but always according to a precise logic. Thus, while all the other prestigious showrooms offered large showcases with numerous watches dis-

Extra-thin gold case and champagne dial for this ladies' model whose design is ahead of its time. Note the long, straight lugs and the applied gold chapters. 9P movement (1957).

The case in white gold and the mirror-plated dial distinguish this elegant model for men. In order to reduce the thickness as much as possible, the hour markings were painted directly on the dial. 9P movement (1960).

Extreme elegance combined with sporty aspect: those are the two "souls" of the Piaget production. These concepts emerge clearly from these two pages of advertising taken from 1940s publications. In particular, the advertisement that emphasizes the importance of the watch for airplane pilots and frequent travelers looks very modern.

played. Valentin Piaget had smaller cases made that presented small groups of watches. Looking inside, neither watches nor jewels could be seen; instead, there were small armchairs, elegant small tables, sofas, and carts with drawers in which watches and jewels were presented. Intrigued, the client entered, to be received with due ceremony and offered a seat. "Thus arrived," recalled Emil Keller, "we could show him all the watches we wanted calmly and in the way we wished to, even if he had come in looking only for a pair of cuff links. We would serve the best coffee and Geneva's best chocolates, specially made for Piaget, we would offer the client our catalogue: all of this created a rapport that, in the majority of cases, ended in a sale, and that's how we built our reputation." After the creation of an advanced facility and a demonstration of technical mastery with the new caliber, which, in its turn, permitted an original design and a veritable revolution in the realm of ladies' watches, the showroom in Geneva added the last element missing in the picture: the image. All was now ready for the great expansion of the 1960s.

Three examples of watches set into old gold coins. The open watch is housed in a twenty-dollar coin dating from 1894 (1957).

Above: Other examples of Piaget's advertising from the 1940s and 1950s under the standards of "Luxe" and "Précision". Next to the wristwatches appear the small masterpieces of the coin watches.

Below: This advertisement demonstrates the synthesis of early 1940s style. For men, a classic round watch; for ladies, a narrow watch with gold bracelet.

One of the earliest models with automatic winding and a square case. Note the convex dial and the central second hand (1955).

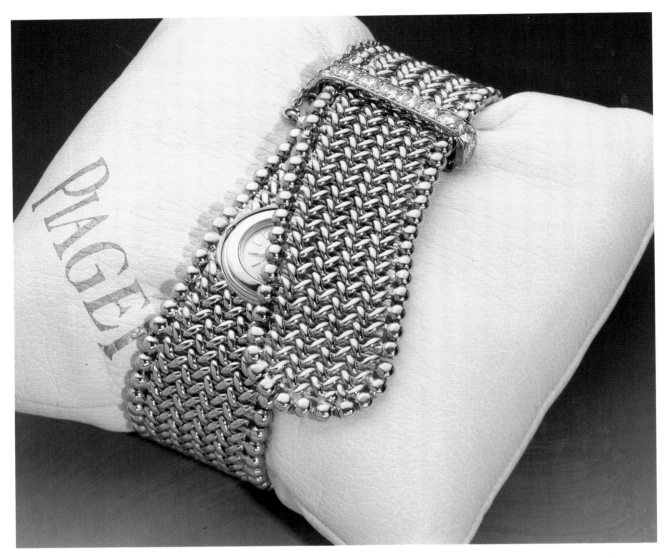

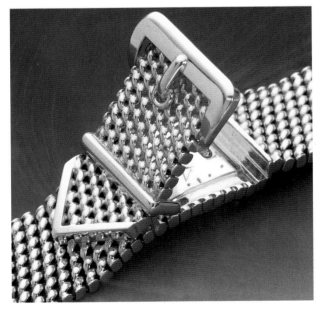

On this page: Two refined wristwatches distinguished by their concealed dial. Above, the model "Manchette" in gold, the clasp embellished with 10 diamonds. 6P movement (1957). Below, the model "Ceinture," whose buckle forms the cover of the dial. 5P movement (1950).

Opposite: An original model for ladies with small triangular lugs, each decorated with three diamonds. The bezel is quite special, engine-turned with a "sun" pattern and four insets of lustrous gold. The movement is a 6N caliber, distinguished by the crown placed on the back (1955).

On this page: One of the most extraordinary models ever realized by Piaget. It is called "Grande Sonnerie," although it is not a watch of that type but a model with minute repeater. In these two plates, the complex movement is seen from the back and the front.

Opposite: On the dial side is seen the button that activates the bell, which "repeats" the hours, the quarters, and the minutes marked at that moment on the dial. The movement is a caliber from 1910, of 12 lignes and 30 rubies, and the watch was made in one model only (1955).

Opposite: A group of watch-jewels for ladies equipped with the 6N caliber. The rectangular model has an oval, black dial (1959).

On this page, on top, from left to right: Gérald E. Piaget and Valentin Piaget. Below, left to right: Edouard, John, William, and Timothée Piaget.

Another superb watch-jewel in pink gold, whose bracelet has been given large links of the "gourmette" variety. The two diamonds were added after the watch was made (1945).

*Above: A group of models from the 1940s.
Left: A watch with manual winding from 1948; beside it is an automatic from 1949.
Below: Two automatic models with date display, the one on the left having a steel case, the one on the right with a pink gold case. The two models (from 1948 and 1947, respectively) are water-resistant and have a push button (at the eighth hour) to correct the date.
Below: A very rare model in steel produced for the American market and distinguished, aside from its military-style dial and phosphorescent figures and hands, by the date display, formed by the center needle with the red tip (1941).*

Above: One of the first automatic models for ladies. The gold case is water-resistant and has a screwed back. Note the large crown and the central second counter. The movement is a 7 3/4 lignes caliber (1956).
Above left: Another automatic model for ladies, but with larger dimensions than the preceding model. The movement is 9 1/4 lignes. Screwed steel back, gold-plated case (1947).

Above right: A model for men from 1945 with steel case having fluted lugs.
Below: Another model with the same movement as the preceding (a caliber of 10 1/2 lignes) but with a gold-plated case and steel back (1948).

A model from 1959, original and futuristic. The watch, with a gold case and a 6N caliber, is distinguished by an interchangeable bezel (available in six colors) and band, also colored, which can easily be changed by means of the small, removable lugs.

Above left: A classic square model with defined edges, having a pink gold case, a gray dial with "cornices" and four gold hour markings, and an engine-turned "sun-rays" bezel. 9P movement (1958).
Below left: An elegant model for men with case and bracelet in yellow and white gold. The black, oval dial has large, gold hour markings. 9P movement (1959).

Above right: Another square model, this one with case in white gold, set apart by the "shantung"-pattern bezel and dial. The hour markings are painted. 9P movement (1957).
Below right: This rectangular model with large, satin-finish bezel, exploits the elegant contrast between yellow and white gold. Silvery gold dial with hour markings in yellow gold. 9P movement (1960).

A Piaget bracelet watch adorns the wrist of Abbe Lane, the famous actress and singer, shown here with her husband, musician Xavier Cugat, and the inevitable small dogs, at the beginning of a singing tour in Italy during the 1960s. (Photo: Angelo Cozzi, courtesy Grazia).

One of the typical ultra-thin models for ladies that greatly contributed to making Piaget world-famous as a fine watchmaker. The gold case has a bezel studded with 48 diamonds. 9P movement (1958).

On this page, left: A square model for men with chased bezel bearing the same pattern as the bracelet. Gold dial with hour markings in "beaded minutes." 9P movement (1958). Right: A ladies watch in which the bezel and dial reproduce the design of the bracelet. 6N movement (1960).

Opposite: A very elegant rectangular model with white gold case slightly curved to better fit the wrist. The white gold dial has painted hour markings. Note the small crown of the winding mechanism, partially inset in the case so as not to disturb the purity of line. 9P movement (1957).

A life of jazz: The great horn player Louis Armstrong, the musician who perhaps more than any other forged the success of the musical form born in the American "Deep South."

OLD OR NEW, THE QUALITY REMAINS CONSTANT

Computer screens that show all the stages in the fabrication of each piece on command and, next to them, the tools and instruments not very different from those used by watchmaking pioneers over three centuries: these are the two faces of the same reality. In La Côte-aux-Fées, unlike in other businesses, the silent work of modern electronics has not relegated the noisy old mechanical machines and tools to the attic, for they remain equally important. The encounter between old and new technology has not produced a victor, as the most important element is something else: the human. It is a human hand that must control each phase of production and that gives its definitive imprint to each part of the watch, however tiny it may be. The Piaget company thus becomes one of the rare manufacturers, in the literal sense of the word – make by hand – active in Switzerland today. But what do we mean by manufacturer? It signifies a quality of performance immeasurably superior to the usual standards and, especially, that each essential component of the watch has been conceived, realized, meticulously assembled in the heart of the company. This involves not simply putting together pieces visualized and produced in other factories but a profound knowledge of the art of measuring time and the mechanics of precision. All those things that are neither invented nor acquired belong to tradition – particularly in the world of watchmaking – which is to say, the heritage that pertains to every company. And, in the case of Piaget, that tradition goes back 120 years; a visit to the building at La Côte-aux-Fées where the movements are manufactured proves the point. Of the 206 people working there, only 140 devote themselves expressly to production; working beside them, the master watchmakers of tomorrow serve their apprenticeship. Piaget is one of the few companies authorized to train specialized workers. Actually, the young apprentices number only 8; after four years of courses, they will enter the world of watchmaking. Nevertheless, those who choose to remain at Piaget must undergo an additional year of apprenticeship in order to learn to work on the company's calibers following

From drawing to reality: a jeweler assembles the setting of a watch bracelet that will become part of a complete parure.

Expert hands guide the scooping out of "dimples" in a gold dial to receive diamonds, which will cover the dial, and which might number up to 360. The work is accomplished by means of a small electric drill.

specific standards of quality. At about six o'clock in the morning, the machines will begin to hum, which is another aspect of the tradition, for it is important to benefit from as many hours of daylight as possible. The fabrication of a movement begins with a bar or thin plate of brass, which is first cut to obtain the basic piece that will then be submitted to a series of other operations, manual as well as on machines, before it takes shape as the finished part, ready for assemblage. It takes dozens and dozens of these operations, after which it looks entirely different from the way it began. The main plate of the 9P caliber (that is, the small plate that constitutes the skeleton of the movement and on which all the other parts are mounted), for example, goes through sixteen different operations – correction of its proportions, the drilling of holes, the scraping of burrs – some done by machine, others entirely by hand by specialized workers. And that is only the beginning, for the following operations, all essential to bring to fruition the famous extra-thin movement with manual winding, number close to 630. The manufacturing of the 12P is even more involved, as it is an automatic movement whose construction is more complex: 874 different operations, and only four hours to assemble and fine-tune. Taking a tour of the factory, you would go through a series of rooms, small and large, with enormous windows, and in each of these see adroit hands putting the final touches to tiny masterpieces. In one of these rooms, for example, small tools, such as pliers or minuscule screwdrivers, are fashioned on old lathes. This also belongs to the tradition of Piaget, which has always preferred to make whatever it could "in the house." In another room, a woman is bent over a shiny small disk – the dial of a jewel-watch in white gold – delicately hollowing out minuscule niches in which will be set diamonds that will cover the entire object. She works with a sort of drill activated by a rubber pulley: she carefully lowers the point to dig the "dimple" in the metal where the diamond will be placed. For the larger dials of models for men, she will repeat this operation 360 times, that being the number of stones to be mounted, while around her accumulate the tiny curls of gold that will be gathered and remelted. Several feet away, in another room, is proof that very high-quality quartz watches have taken their place in the manufacturing tradition. For these, except for some elements, such as batteries, Piaget again makes everything itself, including the micromotors, step by step, down to the coils. The latter are made entirely by hand, by coiling around a tiny pinion yards and yards of copper wire four or five times thinner than a hair.

In other rooms, on the other hand, new and old technologies coexist. Next to presses dating to the 1950s (they function perfectly, and it would be absurd to change them, given Piaget's limited production: no more than 20,000 watches a year) sit extremely sophisticated machines. Directed by computer, they can carry out in series a group of complex operations. For example, they could form and dig out in the side of the case the hole through which the stem of the crown will pass. They work with a minimal margin of error, amounting to some hundredths of a millimeter, and are thus very precise. Nevertheless, the ultimate control of their work is human, working by means of microscopes and other instruments of verification that would allow the scrapping of those pieces not up to the requisite quality. Another fascinating moment, which might help to give an idea of what the word manufacture means, is the "signature" of the dials. For this, as well, a specialized technician performs the operation: he inks by instants a metallic plate on which the logo Piaget is engraved; on the plate, he applies a rubber stamp in order to transfer the inscription, then lowers the

The finishing touches are given to the bezel of a "Polo" watch before the final polishing. Through artisan's mastery, all unevenness and imperfection is eliminated.

stamp until it just touches the dial. It is an operation requiring both firmness and delicacy, and it may have to be repeated several times, as the slightest problem, a vibration of the floor or a particle of dust, could result in the name not being perfectly printed. The most intriguing room of all is the one in which the movement is assembled and placed inside the cases and the bracelets that arrive from the subsidiary company Prodor. No more brass can be seen: bridges and plates have been covered with rhodium in order to counteract any problems with oxidation, and their surfaces have been circular-grained in "partridge's-eyes" or decorated in "côtes de Genève." The angles of the bridges and plates have been worked by hand to give them a brilliant polish; the balances have been equilibrated; each moving element has been oiled. This room

This watchmaker is in the process of assembling the various components of the 9P caliber. The tolerance is of some hundredths of a millimeter, and the entire movement takes over two hours to put together.

One of the most delicate operations of the assembly: the placing of the balance and the balance-spring.

shows true master watchmakers at work, those who "give life" to each watch and supervise its first steps. On the other hand, the movements begin a round of "tortures": in this final phase, they are submitted to all manner of quality controls before being authorized to go on the market. These range from tests of water-resistance, involving trials at different pressures, to temperature tests, in which the watches go from the Siberian extreme of 40 degrees below zero Centigrade to the tropical heat of 60 degrees. Then there is the most "technical" test, of the watch's precise functioning. Each watch is observed in all positions by means of a machine that measures almost immediately the "beat" of the balance and prints the results.

It is a kind of electrocardiogram, thanks to which the master watchmaker knows what to do to prevent future errors. Finally, a last control – visual – at the moment of placing the watch in its package, ready to be sold.

The computer is never wrong: this electronic apparatus is testing the duration of the winding and the regular unwinding of the mainspring of the watch, equipped with a 9P movement.

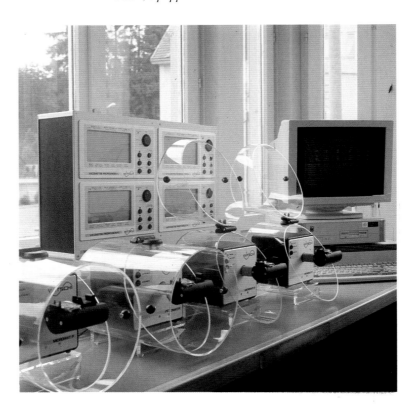

1960 · 1970

The 1960s saw a veritable creative explosion at Piaget. The ultra-thin movements, either manual or automatic, allowed expressions of aesthetic virtuosity unthinkable in the preceding decade. The company's range of styles grew by scores, all of them original and high-class. It was, however, perhaps the watches with gemstone dials that, more than any other, synthesized the new avenues that Piaget was exploring: small masterpieces of the craft of jewelry, good taste, and technology.

The Boom Years

Memorable years, the 1960s. The images, the colors, the music of the time all possess great force: from Pop art to the Beatles, from European cinema and its stars to the explosion of ready-to-wear, from the triumph of plastic to May 1968 – the shock waves of the decade are still making themselves felt. One reason may be that one of the most famous slogans of the student movement in Europe, "power to the imagination," although it never got beyond a utopian idea on the political plane, was fully realized, in many different domains, in everyday life. From the mass market to luxury products, it came out on top, and all that it carried within it left the mark of a creative spirit that opened wide new horizons in art, fashion, design.

In the closed and tradition-bound world of watchmaking, where reputations are determined by a company's history rather than its current performance and are difficult to change or open to question, Piaget proved a truly new and stimulating protagonist from all points of view.

Already in 1957, the decision announced to the public to produce watches only in precious metals, gold or platinum, smacked of a challenge. What was this? The major companies of Rolex, Audemars Piguet, Vacheron & Constantin, Patek Philippe, Jaeger Le Coultre, Omega, and many others much better known than the almost unknown Piaget did not disdain steel even for such important time-pieces as the chronographs or the astronomical watches, while this parvenu allowed itself this aristocratic snobbism? Yet Piaget's decision was perfectly in keeping with its advertising messages ("luxe et précision", "the watch of the international elite"), as well as its proclaimed vocation of watchmaker-jeweler.

As for the creative spirit typical of the 1960s, Piaget

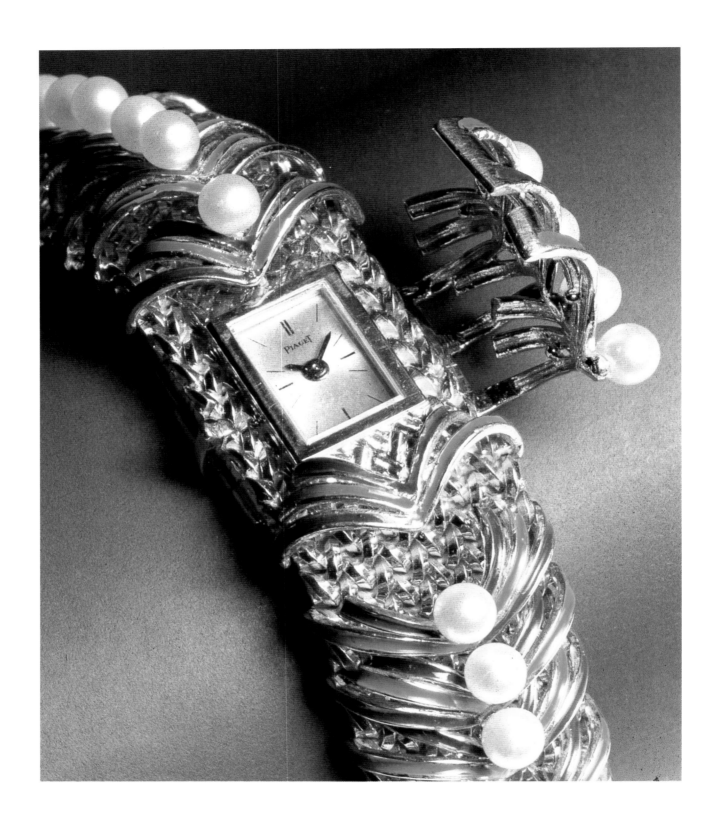

Bracelet watch with concealed dial. This model in gold and enamel is embellished with 38 pearls. The baguette-shaped 2P-caliber mechanical movement has the crown on the back (1968).

Hunter pocket watch with white gold case entirely set with small diamonds (1,094 stones with a total of 13.15 carats). 9P movement (1964).

started off right away with a resounding technological innovation. After the classic ultra-thin 9P movement, presented at the Basel Fair in 1957, appeared in spring 1960 the legendary 12P caliber, which, at only 2.3 millimeters thick, was long considered the thinnest automatic movement in the world. The rotor was made in 24-karat gold to guarantee maximum pressure, while thirty rubies assured the perfection and precision of its movement, which had a power-reserve of thirty-six hours. The first technical drawings were adjusted by Valentin Piaget in August 1956, and the patent was registered in May 1958 (no. 339571) with the Federal Bureau of Intellectual Property under the name of Complications S.A. This was the company, with headquarters in La Côte-aux-Fées, in which Piaget grouped its research and development departments as well as the manufacture of movements since 1948.

Piaget's display at the Basel Fair of 1960 was, as usual, small, but this time it became the object of a lively curiosity and a steady stream of technicians and heads of rival firms (some of whom refused to believe that such slim watches could house automatic movements). The influential *Journal de Genève* characterized the birth of Piaget's 12P caliber as "an event destined to enter the annals of horological history."

And so it did. Piaget's designers, having at their disposition this amazing ultra-thin mechanism, used it primarily in several of the most classic models, which already inspired the inimitable look of the 1960s. Besides the variations on the basic automatic model, there was the automatic calendar, in its standard version or with water-resistant case (another example of great technical skill, given the slender dimensions). Or model 12103, with round gold case, available with different dials (in 1961, this cost the then high price of 2,810 Swiss francs); model 12603, slightly

Bracelet watch in yellow and white gold distinguished by its double bezel: one set with diamonds, the other with emeralds. 6N1 movement (1961).

Another model in yellow and white gold, with its bezel in the form of a flower in diamonds and sapphires. Engraved gold dial. 6N movement (1970).

This bracelet watch, whose bezel is ornamented with 24 diamonds, features interlaced floral motifs in white, yellow, and pink gold. 6N1 movement (1965).

Bezel, dial, and bracelet in the same pattern characterize this bracelet watch with gold hour markings. 6N movement (1962).

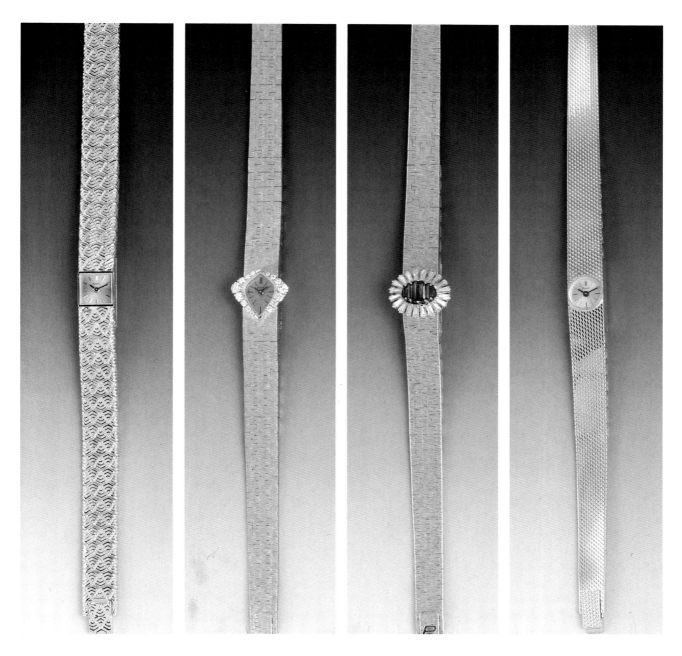

This small bracelet is only as wide as the dial, all in gold, with gold hour markings. 2P baguette-shaped movement (1966).

Rhomboid-shaped bezel set with diamonds on this model in white gold, with bracelet and dial in the same pattern. 2P movement (1964).

Bracelet watch in white gold with concealed dial; the cover is set with baguette-cut diamonds and sapphires. 2P movement (1969).

This gold bracelet watch with tiny dial has painted hour markings and is fitted with the diminutive 2P movement (1963).

smaller, with white gold case; a watch with a typical Piaget shape, the curved square, almost octagonal in effect; yet another version, in white gold, having a dial engraved with the same design as the case; and another with white gold case and dial in mother-of-pearl.

Elitist by nature, the 12P automatic movement was set into very few models in comparison with the 9P, which, since its appearance, had always taken up the lion's share of the catalogue. Considering that Piaget's total production in the boom year of 1966 reached only 10,000 pieces, it is easy to understand why the ultra-thin automatic Piaget watches of the 1960s are today highly sought after by collectors who are not content to follow current fashion.

Meanwhile, with its acquisition in Geneva of a valuable jewelry workshop and businesses specializing in,

The famous automatic 12P caliber makes its debut. For several years, it was considered the thinnest in the world. Here, it is set inside a bracelet watch for men with a cushion-shaped case, its dial and bracelet engine-turned in the hobnail-pattern, the same pattern given to the bezel as well (1961).

An unusual ring watch, in granulated gold. The golden dial, with gold hour markings, has a diameter of 15.7 mm. It uses a 6N1 movement, with winding and setting crown on the back of the case (1964).

The same case used for the model on the preceding page but in smaller dimensions for this ladies' model, with bezel and dial decorated in the hobnail-pattern. 9P movement (1964).

The hobnail-pattern sets off the bezel and dial of this men's model with rectangular case and gold hour markings alternating with painted ones. 9P movement (1962).

among others, the manufacturing of cases and bracelets, Piaget gained total autonomy of production. At that time, the growing importance of the work done in Geneva also symbolized the growing curve of the new specialty: in the future, the name Piaget would ally the art of making luxury watches with the art of the jeweler. During the same period, only Cartier, from the height of the first rank that it incontestably occupied in the realm of watchmaking, could achieve a like synthesis of the two vocations. Who would have predicted the historic conjunction of the two great watchmaking concerns scarcely twenty years later?

As for Piaget, which had going for it the enthusiasm of youth and the clear head start in the realm of advanced technology, its primacy in the area of the ultra-thin mechanisms was highlighted by the unbounded imagination of its designers and jewelers, who created an authentic interpretation of the spirit of the decade. The flat cases of original form, somewhat exaggerated by dimensions and style, stunningly complemented the precious stones that embellished them. These were used not only for unique models commissioned by superwealthy collectors, for which Camille Pilet, Emil Keller, and, from 1966, Yves G. Piaget himself visited the Middle and Far East, but also for regular series models, although in limited editions, that sparkled with rubies, emeralds, and diamonds. Such special pieces loomed large in the catalogue, filled as it was with a great variety of watches for men and ladies (already over 500 in 1963) and emphasizing the high end in any case.

Of particular importance in Piaget's climb to the top was its "conquest," thanks to the excellence of its precious watches, of the most celebrated jewelry establishments in the world: in London, Garrard, jewelers to the queen, and Asprey, furnishers "by appointment" of the Royal Family; in Paris, after an

This white gold model is just an example for the countless high-jewelry developments achieved in the collection of ultra-thin ladies' watches. Sapphires and diamonds surround the lapis lazuli dial and frame part of the bracelet. 9P movement (1966).

A typical jewel-watch for ladies from the 1960s, with bracelet and case in white gold and bezel set with 40 diamonds. Four more diamonds are used as hour symbols on the onyx dial. 9P movement (1966).

The bezel of this model with yellow gold bracelet and hands is set with 32 diamonds and 4 emeralds, while the dial is made of jade. 9P movement (1965).

Another jewel-watch, this one with case and bracelet (designed in a wave pattern) in white gold. The oval dial is lapis lazuli and the bezel is embellished by 28 diamonds. 9P movement (1967).

Two jewel-watches in white gold. The watch on the left is surrounded with 24 diamonds and 28 sapphires. 2P movement (1970). The larger model on the right has 36 diamonds and 60 baguette-cut rubies. The dial is mother-of-pearl. 9P movement (1966).

Above, left: Another model in white gold, set with 28 baguette-cut rubies and 36 diamonds. 9P movement (1963).
Right: A different composition of rubies, diamonds, and white gold, with the dial having the same pattern as the bracelet. 9P movement (1966).

extensive series of visits. Cartier itself, "king of jewelers and jewelers of kings," on the rue de la Paix, finally agreed to sell certain models; in New York, the legendary Tiffany, on Fifth Avenue. It also found prestigious outlets in Madrid, Düsseldorf, Buenos Aires, and Hong Kong. At the same time, owners of high-quality jewelry and watch stores (such as the established boutiques of Fiumi and Pisa in Milan and Monetti and Hausmann in Rome in the important Italian market) passed along to Piaget requests from their clientele, which had become instantly infatuated with square models with precious stones and ultra-thin coin watches. Watches from the series "Hommes d'affaires," for day as well as night, "Tradition," which remain in the catalogue today, as timely as ever, and "Protocole," all both personalized and above fashion, began to be appreciated and sought after.

This extremely refined pocket watch for evening has a white gold case, pendant hollowed in the bezel, and crown on the bottom. Four diamond hour markings stand out against the onyx dial. 9P movement (1962).

Left: A model containing the Beta 21 quartz movement. Note the large dimensions (33 mm wide) and the stepped case, as well as the lapis lazuli dial and bracelet in cylindrical links (1970). Below: Automatic model with cushion-shaped case and black dial with gold hour markings. 12P movement (1961).

So it came about that in a short space of time, the Piaget watch, with its quality and unique, innovative appearance (the famous "Piaget look"), became the preferred watch not only of the jet set but also of the upper middle class and, above all, of the personalities that lent their style to the culture of the period: famous writers, theater and movie stars, musicians, citizens of the world. One has only to leaf through the newsweeklies and French and Italian women's magazines from the beginning of the 1960s to see actresses like Gina Lollobrigida or Sophia Loren sporting splendid Piagets on their wrists. On the snow at Gstaad, music-hall greats like Maurice Chevalier and balladeers like Mireille Mathieu wore one, while the queen of mambo and cha-cha-cha, Abbe Lane, put one on for her strolls on the Via Veneto, much appreciated by the paparazzi.

Right: The dial of this oval watch is made of turquoise; the bezel is engine-turned in the "clous de Paris" pattern. 9P movement (1967).

Below: The dial and outer bezel in malachite distinguish this square model with small lugs and case in yellow gold. 9P movement (1966).

Very original in form, this watch for the pocket or the hand-bag takes the shape of an envelope, with chased gold case. It opens with a light pressure on the center. 9P movement (1965).
Opposite: Two pocket watches in "shutter" form. Pressing the sides of the case opens the two shutters, which cover the dial. The watch above has a case in white gold and the one below is set with 1,344 diamonds. Both models use a 9P movement and date from 1963.

This refined watch has a square case with cut corners and similarly patterned bezel and dial. Its classic styling gave rise to the current line called "Protocole." 9P movement (about 1963).

The elongated rectangular case is in white gold; the silvery gold dial has painted hour markings. 9P movement (1967).

Another rectangular watch in white gold, with engine-turned dial and bezel. 9P movement (1963).

Model in white gold with contoured bezel; bezel and dial are patterned in "clous de Paris." 9P movement (1963).

The ambassador as well as the main architect of the company's success on the level of international classic elegance, from Palm Beach to the Côte d'Azur, from Rodeo Drive to Deauville, was, from the mid-1960s on, Yves Piaget, son of Gérald, born in 1942 at La Côte-aux-Fées. He was educated in Neuchâtel, first at the high school, then the university, earning a degree as specialized engineer in watchmaking. To complete his studies, he went to the United States, receiving a diploma from the Gemological Institute of America in Santa Monica, California. Back in Geneva, he began working in the family business in 1966, at the age of twenty-five, at the most crucial location for a future president: the headquarters on

Elegant chain bracelet watch with large links. The tiny Roman numerals are painted on the gold dial. 2P movement (1963).

Opposite: A group of jewel-watches from the mid-1960s, all containing the 2P baguette-shaped movement with crown on the back of the case. Left: Model with bracelet set with diamonds, rubies, sapphires, and emeralds. Center: Gold case and bracelet with diamonds and cabochon corals. Right: Blue dial and bracelet with blue and green enamel links.

Another model containing the Beta 21 quartz movement. The large, stepped case is in white gold, and the Roman numerals are painted on the dial. The "leaf"-style hands are unusual on Piaget watches (1969).

Opposite: The photograph of a watch surrounded by large quartz crystals underlines the importance of the new method of measuring time. The stepped gold case is again made large to accommodate the new movement (1969).

the rue du Rhône, where design and production, clientele and distribution came together. "Remember that you come here not because of the name you carry but by virtue of your vocation and your training." This is how Yves G. Piaget recalls his father's advice and the initial conditions of his employment, with commensurate salary – just enough to cover the rent of a studio in Geneva ("My wife, who was a teacher, made more than I did," he recalls, "and in fact it was she who enabled us to buy our first car, a Fiat 500").

In Geneva, Yves, the foremost representative of the fourth generation of Piagets, learned the arts of selling and of international public relations, for which a considerable knowledge of languages, people, and, above all, watches, had predisposed him. However, he did not neglect to maintain a close relationship with his uncle Valentin and the world of the factory at La Côte-aux-Fées, nor to keep up his visits to the subsidiary company Prodor, which took in gold, platinum, and precious stones as its raw materials. It could be considered a sort of "gold mine," the kingdom of Piaget jewelry and creativity.

In 1968, Yves G. Piaget was named director of marketing and communications. He began travelling all over the world to launch Piaget's image on an institutional level, by means of a series of campaigns, initiatives, and sponsorships, pursuing in particular the markets with the greatest competition: Europe, the United States, and Japan. Meanwhile, Camille Pilet, who henceforth spent at most two months a year in Geneva, and Emil Keller, handled exceptional clients and unique pieces of inestimable value: watches commissioned by Arab emirs, Indian princes, or collectors with unlimited budgets, such as the Japanese man for whom Piaget would later realize the Piaget "Phoebus," thus named for its splendid stones. But the god of light in Greek mythology was also the

patron of the fine arts, so for Piaget, the name carries a double symbolism.

While this small group of young men filled with energy, enthusiasm, and confidence in the supremacy of their company's watches led the battles in Geneva and the rest of the world, Gérald and Valentin pursued a strategy of expansion, always keeping in mind the rustic maxim never to overestimate one's strength. Their first important step was the acquisition, effected in 1964, of Baume & Mercier. This company, founded in 1830, had gained a solid reputation in the 1920s and 1930s for its shaped watches, its ladies' models, and its jewel-watches. In 1958, the heirs of both founding families had ceded their shares to Marc Beuchat, and Baume & Mercier began to produce splendid chronographs and models with automatic movements. This was a complementary type of production that, along with sports watches, Piaget had never explored. On the other hand, its line of ladies' models, jewel-watches, and ultra-thin automatic watches would add a useful synergy.

Another revolutionary novelty of 1964 was the launching of dials made of gemstones: jade, coral, lapis lazuli, malachite, tiger-eye, opal, turquoise, and many others, the catalogue enumerating more than thirty different stones. Never before in the history of watchmaking had the dial come to resemble the palette of a painter who drew directly on invention and motifs dear to nature. The splendor of these stones, paired with cases of oval form (most oriented on the horizontal, an "oversize" design typical of Piaget) or square, curved square, rectangular, hexagonal, barrel-shaped, even round, enhanced by all the nuances of gold or platinum, made each model a unique collectible. One of the most important and complete collections, assembled by Burton E. Grossman, among Mexico's most prominent industrialists, contained about sixty such pieces, some of them unique, created specially for him.

The newest feat of making the dials in gemstones was aided by the ultra-thin 9P and 12P movements, while in Geneva, Piaget's jeweler artisans miraculously managed to keep the stones used at a thickness of seven-tenths of a millimeter.

With the appearance of models with stone dials in the Piaget catalogues starting in 1964, the Maison Piaget was definitively extolled as the most creative of the great companies. Certainly, it was the one most consonant with the taste, colors, and design of the 1960s, such as an article of clothing by Yves Saint-Laurent in the style of Mondrian, a miniskirt by Mary Quant worn by Twiggy, a Citroën DS 19, or a painting by Roy Lichtenstein.

It was, in fact, the tremendous financial success enjoyed by Piaget watches in the boom years that cushioned the company during the Swiss watchmaking crisis of the 1970s, allowing it to face it with equanimity. Still, the historian who seeks to quantify the success of the company (astonishing, even exceptional, that such revolutionary and fashionable watches were created not in London or Paris or New York but in a tiny village lost in the Jura Mountains) in terms of sales figures, increased share of the market, and other data comes up against one of the numerous rustic habits of the Piaget family. It seems that, while the company was conducted very democratically, while confidence reigned overall, while decisions that affected even aesthetic details or the cost of a new product line were arrived at collegially, the only form of annual statistics kept were on tiny pages from a notebook that Valentin showed to the other members of the family at the end of the year. "A detail that, seen from the outside, might seem strange," recalles Yves G. Piaget, "but which, on the contrary, explains

perfectly the company's philosophy is that none of the Piagets of the third generation had put aside an example of each model produced, in order to create a collection. When I asked about this, I was told that the watches had been made to be sold, down to the last one, so that it was up to me to start a collection in the 1980s, directing Emil Keller to buy back the most significant pieces from clients, collectors, or auction houses."

Toward the end of the 1960s, a new element – the quartz movement – threw the worldwide economy of watchmaking into a tailspin. The first models that carried it, in 1968 and 1969, cost a thousand dollars, but it was not long before it became affordable, as soon as the initial expenses of research and production were amortized. It was easily reproducible in big quantities and proved much more precise than any mechanical movement. On the surface, it seemed that, thanks to this invention, whoever had enough money could quietly begin to make watches (which is exactly what a multinational like Gillette, for example, did). It also seemed to signal the beginning of the end for Swiss watchmaking, with all its traditions, its centuries of history, its heritage, and its master watchmakers. However, this did not come to pass, although many distinguished names in watchmaking floundered before being rescued by a consortium of Swiss banks – but that is another story.

Preceding page: The malachite dial of a model with oval case. Note the veins of the stone and the perfect Piaget signature on the dial. 9P movement (1967).

On this page, left: Gold ladies' model with 2P movement (1963). Next to it is one of Piaget's best-sellers from those years. The 23-millimeter-wide watch (not including the crown) has applied gold hour markings as well as black painted strokes and a 9P movement (1961).

O ther great companies, including Piaget, reacted in a more intelligent manner, trying neither to exorcise the new discovery nor fight it. Instead, Piaget sought to assimilate the technology, introducing it partially into its production, following market demand, while continuing to produce valuable mechanical movements and taking care, above all, not to change the company's image. Thus, the Centre Electronique Horloger (CEH) came into being in 1962 at Neuchâtel. This was a sophisticated research center focused on the new electronic technologies, to which Rolex, Omega, Bulova, Longines, Jaeger Le

The extra-thin 9P movement runs this ladies' model, with similarly patterned dial and bracelet. The bezel is set with 30 diamonds (1962).

Cushion-shaped case and automatic movement with date display characterize this gold model with applied hour markings. 12PC movement (1965).

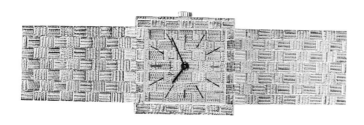

Bracelet, dial, and bezel all carry the same pattern on this elegant gold watch outfitted with a 9P movement (1963).

Coultre, Elgin, and IWC, as well as Complications S.A. (Piaget), belonged. The quartz movement that Piaget introduced from the end of 1969, after two years of work at CEH, lost precisely ten seconds in the space of eight months. Of course, these movements were relatively large, which constrained the stylists of the firm and forced them into a sort of design acrobatics. But during the 1970s, once the technological problems, and with them the aesthetic problems, had been resolved, quartz watches made their appearance in the catalogue, alongside the mechanical watches that had once been the most innovative, true to the spirit of Piaget.

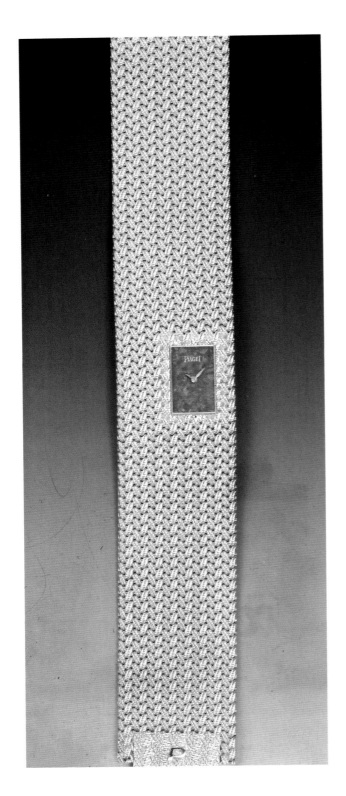

The opal dial is placed off center and has no hour markings in this bracelet watch with tight-fitting links of white gold. 6N movement (1970).

Another ladies' model in white gold with a caliber 6N. The round dial with applied hour markings is set off by 4 marquise-cut diamonds and 16 baguette-cut sapphires (1961).

This exceptional pocket watch has a detachable châtelaine. The white gold case has pavé-set diamonds. Four more diamonds on the onyx dial constitute the hour markings. This model was also made with baguette-cut rubies, sapphires, and emeralds. 9P movement (1962).

Two medals, commemorating Winston Churchill (1965) and Pope John XXIII (1968), enclose watches with the 9P caliber. Below them is a coin watch made from a Mexican gold twenty-peso piece, with the 4P caliber (1965).

Opposite: Two refined models with chased bezel carrying the hours in Roman numerals. 9P movement (1966).

Even cuff links can turn into watches, which Piaget realized in this novel creation. 6N movement (about 1960).

A classic round shape graces this extra-thin model in white gold, with engine-turned bezel in the "clous de Paris" design, small, contoured lugs, and white dial with extremely slender painted hour markings. 9P movement (1963).

These three round models of the 1960s epitomize the classic elegance of all Piaget watches. Above: Gold dial with Roman numerals. 9P movement.
Center: White gold case, silvery gold dial, and automatic 12P movement. Below: Ladies' model with white dial, 9P movement.

A variation on the watch with classic shape, with engine-turned bezel in the barleycorn design and malachite dial. 9P movement (1965).

Bracelet and bezel are engraved with the same pattern on this extra-thin watch with automatic 12PC movement and date display (1967).

Another automatic model, its bezel patterned in "clous de Paris" and metallic blue dial with gold Roman numerals. 12PC movement (1964).

This watch from 1960 is among the first extra-thin models carrying the automatic 12P movement (2.3 mm thick). It has an extremely smooth gold case, large bezel, and small lugs. The dial has applied hour markings in gold.

A meeting of two celebrities, both wearing Piaget watches on their wrists. Maurice Chevalier, the great singer, appears to be giving advice to Mireille Mathieu, one of the leading singers of the 1960s. Beside them is Yves G. Piaget.

Yet another interpretation by Piaget on the theme of the shaped watch (that is, not round). The rectangular case does not reveal its attachment to the strap, while the dial features painted Roman numerals and a "railroad" minute track. 9P movement (1963).

This extremely elegant watch has a gold barrel-shaped case that seems to form a frame for the engine-turned onyx dial. 9P movement (1969).

The Beatles, the most famous quartet in the world, symbolize the music of the 1960s. Here, the four boys from Liverpool, soon to become baronets, take a walk in London in 1965.

TIME
AND THE JEWEL

Who said that all that glitters is not gold? At Prodor, this famous adage does not hold water, for here all that glitters is truly made of gold, or, even more precious in this place where fine jewelry sets and watches are made, of platinum. In this Geneva factory, created by Piaget to specialize in the making of watch cases, bracelets, and jewels, many pounds of these costly materials undergo a metamorphosis every week. Even the air inside the building is precious, filled as it is with infinitesimal particles, the microscopic residue of operations on the metal, which, although invisible to the naked eye, settles on the clothes and bodies of everyone in the workshops. On entering, a visitor might mistake it for a clinic rather than a watchmaking workshop, for all the artisans wear smocks, and visitors are asked to put transparent plastic guards over their shoes. The explanation is not arcane: nothing leaves Prodor before it is washed with care, or perhaps burned, then filtered to recuperate the particles of precious metal. Besides the prodigy of the enterprise itself, with its modern machines,

A watch-jewel with concealed dial, in white gold, diamonds, and baguette-cut sapphires. When the sapphire placed at the sixth hour is pressed, the cover opens and reveals the dial. 2P movement (1972).

technicians, designers, jewelers, and gemologists, the system of air purification and water filtration constitutes another technological marvel. By means of this system, many pounds of gold are recovered each month, which is melted and then reintroduced into the cycle of production.

If production at La Côte-aux-Fées begins with a plate of brass, at Prodor the day begins with a gold bar or long wire. The former serves as the basis for a watch case, while the latter represents the first step toward a bracelet. In general, the case of a bracelet watch has three parts: the caseband which forms the supporting structure in the middle; the back, which is fixed to the caseband and closes the watch on its rear; and the bezel, which keeps in place the crystal that protects the dial. On certain models, the case has only two parts: the back and the caseband, which fills the function of the bezel as well. Nonetheless, despite the small number of parts, hundreds of operations must be carried out to complete a Piaget watch case. Such is the road to absolute perfection, which begins with a rondelle of gold weighing approximately 90 grams, cut from a bar. It then goes from ultramodern machine to the adept hands of the artisan in a number of operations. After each of the initial phases, the case, which gradually begins to take its definitive shape, is "cooked" in small electric ovens at 700 degrees Centigrade. This process is essential in order to return to it its basic characteristics after undergoing the "stress" of cutting and punching. The technicians call it the "reorganization of molecules," and its purpose is to avoid the advent of microscopic fissures in the heart of the metal that could impair the rigidity of the case. Rigidity is important to the correct functioning of the watch, especially since most of the

Case and bracelet in white gold, entirely covered with diamonds, some of them baguette-cut. The oval dial is made of onyx and has no hour markings. 9P movement (1974).

cases are of the extra-thin variety, and the slightest torsion could affect the movement and impede its reliable operation.

After the major processes of stamping and punching, the cases undergo finer operations by automatic machines under the direction of a computer, although, of course, the computer is always under the human direction – it is a human expert who verifies that the solder of the lugs does not show the least imperfection. Meanwhile, artisans polish and smooth (exclusively by hand) each corner or edge, so that the finished case will feel "soft" to the touch, like silk. Others decorate the back with "partridge's-eye" motifs, or check the threads of the tiny screws, also in gold, that will fasten the back to the caseband so that the case is hermetically sealed. It is left to female hands to carry out the different phases of the final polishing, which precedes the ultimate testing of the cases at every conceivable angle and, finally, their wrapping in vellum paper to avoid scratches. Those rondelles of gold, originally fat, after hundreds of hours of work have become small "strongboxes" destined to protect for a very long time the "time machines" manufactured at La Côte-aux-Fées. At the same time, a very different kind of work unfolds in the workshop: one that will give birth to a bracelet for the watch. This work begins with a tiny oval link, which originated as a long gold thread. For fifty years, the same machine tool has carried out this operation: the gold thread enters one end and the links accumulate with a tinkling sound on the other. With these, Prodor's chain makers create "soft textiles" of gold destined to encircle the wrists of the most refined clients. A Piaget bracelet is none other than a gold textile whose fabrication has much in common with that of a rare carpet. Very fine gold pins constitute the woof, while the design is formed by the links, inserted into the pins one by one, filed in order to fit perfectly one with the next, then partly hooked together and partly welded. As the links slowly collect next to one another, the chain maker takes his small blowpipe and lays the fragment of bracelet manufactured on a wooden cylinder while pressing the fragment against the

One of the extra-thin watches for ladies made by Piaget. It has a lapis lazuli dial, set off by a prominent bezel embellished with diamonds. 9P movement (1976).

cylinder. This process, repeated hundreds of times, lends suppleness and roundness to the bracelet itself, which little by little loses its stiffness and adapts to the wrist just as a textile would. The last two phases are very delicate: the soldering and polishing of the outer edges, which must not present the least roughness, and the soldering of the bracelet to the case, which requires two steps.

First, using a special glue, the bracelet and case are joined and kept in position, then they are soldered together with a small oxyhydrogen blowpipe. However, the work is not yet finished. Hundreds of hours have passed since the chain maker began to assemble links and pins, but many more will pass before the bracelet is polished and shines flawlessly. Here, as well, there is no machine that can carry off this job; only the sensitivity of hands and expert eyes can evaluate the pressure and the angle of attack required when the bracelet is submitted to the action of more or less hard brushes, capable of eliminating any remaining rough traces of the solder and making the gold shine. Obviously, this type of chain bracelet is not the only one made at Prodor. Another is the famous "A" bracelet, whose tiny links, adroitly joined, compose a compact yet supple base; those who do the engraving and engine-turning have come up with almost unlimited motifs with which to embellish its surface. Since its creation in the late 1950s, many other bracelets have been conceived, often characterized by ingenious structures and the subtle joining together of their components, some of which set off beautifully the fire of precious stones.

At Prodor, however, the most intense brightness comes from the top floor of the building. This is where the most singular dreams of clients take shape: the high-jewelry watches. Diamonds, rubies, emeralds, and other precious stones cut in many different shapes line the compartments of strongboxes. Here can be found the master jewelers, engravers, and mounters. To them is entrusted the task of transforming into reality the concepts of watch and fine jewelry designers. Into the warp and woof of the bracelets a third element is introduced: the precious stone. It is difficult work, as all the stones must be perfectly aligned, as well as solidly set into the bracelet to make sure they do not fall out – and all this without sacrificing the softness and pliancy expected of a work of art signed Piaget. If a "normal" bracelet requires hundreds of hours of work, here a single piece can easily take more than a thousand.

This remarkable pocket watch for men belongs to the same family as the one shown on page 109. Decorated with diamonds and rubies, it has a white gold dial with applied hour markings. 9P movement (about 1975).

Although the mount is often invisible once all the precious stones are set, the work involved in accomplishing it is highly complex. But that is not all: the watch case itself, as well as the entire dial, could become the backing for dozens, sometimes hundreds, of stones, which might embellish the bezel or the sides of the case. The results: the celebrated high-jewelry watches which have made Piaget world famous for its ability to join a great watchmaking tradition to the finest expressions of jewelry.

Another superb example of the art of jewelry paired with that of watchmaking. Its oval dial is made of opal, and large baguette-cut diamonds run through the center of the bracelet and case. 9P movement (about 1970).

1970 · 1980

An exceptional example of Piaget's artistic creativity, this bracelet watch from the "slave" collection has an oval dial made of coral, echoed by medallions of the same shape and material in the large bracelet. 9P movement (1971).

High Fashion, High Technology

People in the business affectionately called it "the monster." This was the Beta 21 caliber, the first quartz movement to originate in Switzerland. Piaget, which, as we have seen, belonged to the consortium that created it, sought to make it likable, like Walt Disney's Eta Beta or Steven Spielberg's E.T. It found two ways to accomplish this, thanks to its talented designers. The first followed a trick of the great couturiers: masking a flaw (its excessive volume) by disguising it through the shape of the case (such as the one resembling a television set, and dubbed accordingly). The second was by diverting attention to another component usually relegated to the background, such as the bracelet, which here might come to the forefront by virtue of its precious character and consistency or its workmanship. The first model provided with a Beta 21, for example, had a heavy gold bracelet in horizontal lines that adjusted almost instantly to the client's wrist.

One of Piaget's most original collections among those presented in the early 1970s was that of the cuff watches. Intended for young, up-to-the-minute women, these watches boasted wide dials and bracelets that, once on the wrist, assumed a cylindrical shape. The aim was to offer a singular kind of watch well adapted to the wrist.

Another original collection from the early 1970s was that of the *sautoir* watches, which could be worn around the neck, on chains both precious and a little jazzy, very much in line with the fashion of the period (it is not hard to imagine these worn by models such as Veruschka or Donyale Luna), which combined an ethnic flavor, the psychedelism of flower children, and the neo-romanticism of Laura Ashley and Emmanuelle Khahn. Some of these models could be detached from the necklace and worn at the wrist.

A precious necklace watch in gold with a large dial in turquoise lacking hour markings. The crown is hidden under the bezel. The large "pearl" at the end of the chain is also made of turquoise. 9P movement (1971).

Another example of a "slave" bangle watch, with a rectangular dial in coral and cabochon turquoises (32 in all) studding the bracelet. 9P movement (1971).

Of course, dials made in gemstones, which had contributed largely to establishing Piaget's reputation on an international level since their appearance in 1964, came into use for quartz watches as well. The great novelty of this period, which once more confirmed Piaget's creative genius as a jeweler, was the employment of such stones not only for dials but for the bracelets as well, as much for ladies (lapis lazuli, coral, turquoise, and so on) as for men (onyx, tiger-eye, mother-of-pearl), using geometric motifs for a sparer, very classic look.

Starting in 1976, watches for ladies gained the benefit of the mythical 7P, the thinnest quartz movement in the world, in every price range, from the simplest models for everyday use to the high-jewelry watch for night, with their diamonds and other precious stones. Before the breakthrough of the 7P caliber, quartz watches represented only about 5 percent of the models in the catalogue. "About the war against the Japanese," commented Yves Piaget, "at the top of the line, that is, for Piaget and houses like Patek Philippe, Vacheron & Constantin, Audemars Piguet, and other greats, it did not exist, there could not be a war. Japanese electronic watches made problems for companies producing moderate or lower-priced models but had no effect on the watchmaker-jewelers nor the makers of high-quality, sophisticated mechanical watches."

Leafing through Piaget's catalogues from 1970 to 1974 offers, besides great aesthetic pleasure, the feeling of soaring in a jewelry heaven, in a very special world far removed from the political, social, and economic woes of the period (such as the oil crisis that brought on the 68 percent price hike of crude oil by the oil cartel). Thus, the catalogue of 1970 proposes the return of the necklace watch that embellished the necks of beautiful aristocrats of the eighteenth and nineteenth centuries, recast by the designer Jean-Claude Gueit, Piaget's leading creative mind, in contemporary form perfect for the fashionable discothèques of New York, like Hippopotamus, or exclusive clubs like London's Annabelle and Paris's Régine. The large, extravagant cuff bracelets of 1971 confirmed Piaget's originality in relation to the other great makers of jewel-watches, thanks to a special collection called "the Fauves" (the beasts), in reference to the gemstones used especially on the bracelets and dials, such as tiger-eye. In 1972, the accent shifted to mechanical movements paired with precious and ornamental stones in watches with usually classic forms. This trend continued in 1973. "Inspired by the marvels of nature," we read in the catalogue. "Piaget has sought to extract its most precious treasures in order to offer, after joining them in jewel-watches of unequaled luxury, a homage to beauty."

During the 1970s, the creator of the extra-thin watch concentrated on large cases in round, oval, and octagonal form.

In 1974, the company's catalogue, dedicated to special models or classic watches in the purest Piaget style, displayed more than 1,200 models.

Looking through the catalogues from 1959 to 1988, as well as Piaget's advertising, a consistent sensibility emerges, elegant and very classic. In the realm of communication, credit belongs to Roger A. Dick and his agency, Publi-Conseil of Geneva, for creating this "Piaget look" and making it inimitable.

If it is now clear that in the world of watchmaking Piaget almost alone personified all the values that the haute couture embodied in the world of fashion, it is similarly obvious that the special models discussed

On this page: The gold case and bracelet unify the alternating medallions of onyx (including the dial) and coral. 9P movement (1974).

Opposite: A striking jewel-watch in gold whose large-scale bracelet (41 mm) is fashioned to resemble "tree bark." The dial is in tiger-eye. 9P movement (1970).

Above: Unusual hexagonal model with gold case and "twisted" bezel. 9P movement (1975).
Below: The same hexagone transposed to a bracelet watch with "twisted" bezel and edges, links and dial set with coral and diamonds. 9P movement (1975).

Opposite: Another jewel-watch with large-format cylindrical bracelet (57 mm).
Small "bricks" of lapis lazuli alternate with those of turquoise; the dial is in lapis lazuli. 9P movement (1970).

This jewel-watch from the "slave" collection is made of white gold and set with 1,000 small diamonds that frame the opal dial. The bracelet opens by means of a hinged clasp. 9P movement (1971).

The asymmetrical watch with a lapis lazuli dial is set in a large, gold "slave" bangle. 9P movement (1974).

On this page: White gold jewel-watch with a large bracelet whose links are adorned with 368 small diamonds and alternately inlaid with turquoise and onyx. Dial in onyx. 9P movement (1977).

Opposite: A small, rectangular dial in turquoise is set inside a bracelet composed of "pearls" of turquoise and lapis lazuli. 6N movement (1970).

here up to now (and whose prices, in Swiss francs of the period, ranged from five to one hundred thousand) would find an exceptional clientele, like the leading lights of the international high society, with whom Yves G. Piaget associated regularly and for whom he sponsored a team of polo champions, first in the United States, then in Great Britain and Germany. In 1976, the prestigious Polo World Cup at Palm Beach carried the name Piaget, while in 1989, the celebrated Courses d'Or, sponsored by Piaget, made its debut in Deauville. Richly endowed racecourses for horse racing, they were destined to become the gathering place for sports figures and high society of a very high order. Besides the personal passion Yves harbors for horses, "there is a strong link," he maintains, justifiably, "between the world of great foals and that of great watches; in both cases, the breeders and the artisans produce creatures that reach a peak of beauty and performance." A limited account of the most important awards garnered by Piaget in this period includes the Coupe d'or du bon gout français (Geneva, 1969), the Diamonds International Award, also in 1969, given for a bracelet watch conceived by Jean-Claude Gueit, the Mercure de l'élite européenne (Paris, 1971), the Triomphe de l'excellence européenne (Monte Carlo, 1976), and the Laurier d'or de la création et de l'élégance internationales (Monaco, 1980).

At the same time, the fourth generation of Piagets, in their turn, took on important roles in the company which, following tradition, were very diversified. Of the three sons of Valentin Piaget (that is, Yves's cousins), two worked for the firm: Philippe coordinated production, while Gabriel, an engineer-technician, collaborated on the level of the manufacture of movements as well as research workshops devoted to setting up new systems, an area for which he remains responsible today. Both worked at La Côte-aux-Fées.

Ultra-thin watch with white gold case and hidden attachment to the strap. The alternation of mother-of-pearl and onyx segments covering bezel and dial characterizes this model. 9P movement (1975).

Another men's ultra-thin watch has a hexagonal case, also in white gold. Here, the bands of onyx and mother-of-pearl form a geometric motif on the dial only. 9P movement (1975).

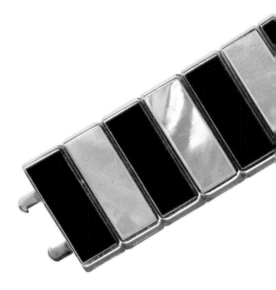

This bracelet watch in white gold alternates "bricks" of mother-of-pearl and onyx. The hexagonal case measures 35 by 28 mm, with a thickness of barely 5.5 mm. 9P movement (1975).

Whereas Yves adored horses, Philippe and Gabriel loved engines – high-performance cars and motorcycles, such as Porsches and Yamahas, with which they took part in races or abandoned themselves to the pleasure of speed during the months when green took over the meadows and the intriguing and deserted roads between the mountains had no snow.

According to all the records, 1974 should have been the Silver Jubilee, Piaget's hundredth year. However, the event that best marked this occasion was the presentation at the Basel fair in 1976 of the 7P caliber as the thinnest quartz movement in the world. It barely measured 3.1 millimeters thick, which meant that the watch, including gold or platinum case and sapphire crystal, had a total thickness of only about 5 millimeters. The caliber possesses an exceptional precision: its daily loss, reinstated by an atomic chronometer, measured in the hundreds of a second. The movement is regulated by a high-frequency quartz (524.288 hertz) particularly imperceptible to the outside world; a microbattery provides it with energy for close to two years.

With the mythical 7P caliber (its code name) quartz movement, the engineers of La Côte-aux-Fées solved a very complicated problem of miniaturization: they succeeded in integrating in a movement 3.1 millimeters thick an electronic module encompassing close to 900 transistors. In addition, they adapted to this module a system that makes it easy to change time zones, by simply shifting the hour hand, without interfering with the precise function of the minute hand. The 7P movement improved another inconvenience of quartz movements: the elimination of the exact time in its memory when, for example, the battery is changed. Piaget's new movement retained an exact memory of the time, which could not be erased.

A single battery fueled all the functions of the quartz movement, thanks to which its memory of the precise time became perpetual. Even replacing the battery would not affect the keeping of precise time.

Once more, Georges Piaget's motto, "Always do better than necessary," which had always been transmitted from father to son, had been honored. A small Swiss company performed valiantly on the preferred field of the Far Eastern producers – electronics. By means of this new technological breakthrough, Piaget consolidated its reputation as a watchmaker ahead of its time and, above all, opened the way to an entire series of new models, while freeing itself from the awkward conditions imposed by the original Beta 21 quartz.

This ended the era of the beast and continued that of the beauty – which had always had the upper hand at Piaget. Starting with the catalogue of 1976, the new 7P showed up in many classic shaped watches for men and ladies, as well as alluring evening models in white gold or platinum. Piaget was gaining a reputation for being the watch of charming international playboys who were to be found on the most beautiful yachts of Portofino or Saint-Tropez or who

Two round models with case in white gold (above) and yellow gold (below) are characterized by bezel and dial in the same material, respectively, tiger-eye and malachite. 9P movement (1975).

Left: Model with octagonal case and oval dial in lapis lazuli. 9P movement (1971).
Opposite: Detail of a watch in mother-of-pearl and onyx.

A large bracelet watch (33 mm) in white gold inlaid with lapis lazuli and jade. 9P movement (1971).

This elegant and supple mesh bracelet shelters a watch with turquoise dial. 6N movement (1970).

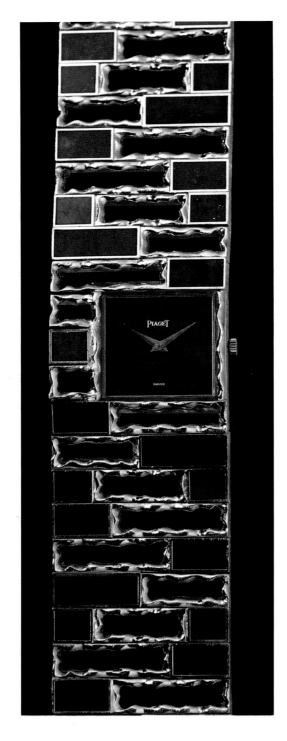

A large bracelet watch (33 mm) in white gold inlaid with lapis lazuli and jade. 9P movement (1971).

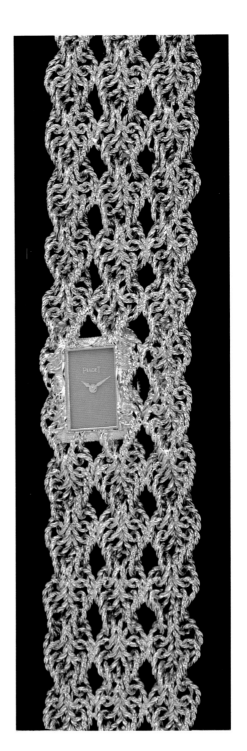

This elegant and supple mesh bracelet shelters a watch with turquoise dial. 6N movement (1970).

A floral-inspired design in white gold, malachite, lapis lazuli, and diamonds. 9P movement (1974).

Jewel-watch in gold, with dial and links in tiger-eye and embellished with diamonds. 9P movement (1975).

This barrel-shaped watch in white gold has an opal dial and a bezel set with 104 diamonds. 9P movement (1975).

One of the models made for the heir to the throne of Bahrain; the dial carries the state's emblem. The case is in white gold and the bezel is decorated with 34 diamonds. 9P movement (1973).

Above: Oval model with double-chain-link bracelet, malachite dial, and gold hour markings. 9P movement (1973). Below: The single hand marking the minutes gives this watch a mysterious aspect. The hour is indicated by the diamond set in an onyx disk. 9P movement (about 1975).

Left: Quartz watch with gold case that returns to the stepped motif of the earliest quartz models with the Beta 21 caliber, but in reduced scale (26 by 30 mm). Dial with Roman numerals. 7P movement (1976).

Below: This model, like the one above, is a quartz watch. Its cushion-shaped case in white gold was also inspired by the first models to carry an electronic "heart." Lapis lazuli dial, with small, silver minute dots. 7P quartz movement (1976).

A round, gold case contrasts with right-angled lugs in this classic design, characterized by a prominent bezel polished like a mirror and a gold dial. 7P quartz movement (1977).

swooped up to the ritziest hotels of Saint-Jean-Cap-Ferrat, Biarritz, or Deauville in their Ferraris or Maseratis; a watch expected as an engagement gift by young and rich heirs; and also the preferred watch of French and American women or European women executives.

In 1979, "Polo" made its appearance, a collection a bit like the squaring of the circle: at once revolutionary and an instant pure classic. It was one of the unique watches that last well beyond a generation. It gained its revolutionary aspect from its seamless look; square or round, for men or for ladies, it is hard to tell where dial leaves off and bracelet begins. The whole had become a single piece, sculpted in gold. Its bolt motif is a masterpiece of precision: each bolt is made separately, to be united to the next with an unimaginably fine exactitude. In work like this, the hand of the specialized artisan proves indispensable, very much in the spirit of one of Yves G. Piaget's favorite expressions, who likes to define his company as a "manufacturer" of watches. "Polo" immediately met with a tremendous success. From its design, which ended the decade with a thing of beauty, to its name (among all the earlier names chosen for the Piaget collections, none etched themselves in the memory of the international clientele so clearly as this), everything connected with "Polo" was perfect – the classic watch for twenty-four hours, an elegant timekeeper which can also very justifiably be described as sporty, due to its solidity and water resistance; perfect for a tennis party or cocktails. Indeed, in the course of its history it would deviate into an endless stream of versions (always with ultra-thin movements, mechanical or quartz), including jewelry models with dial and bolts in white gold and diamonds. Even when a number of Piaget's clients began to demand it with a leather strap, which somewhat counters the conception and the uniqueness of the model, Piaget satisfied them: in these models in solid gold, the horizontal lines became vertical.

Among the original "Polo" collections, we find that the versions most expressive of the Piaget look remain those in white and yellow gold with black dial, in large format for the round watch, in unisex format for the rectangular watch, which, for men, became a superb bracelet that told the time.

The "Polo" line was introduced in 1979. Its immediate international success enhanced once more the Piaget myth. The watch gains its particular look from the integration of bracelet and case, and from the dial that continues the link design. "Polo" was also realized in a square version without losing its inimitable personality, a rare feat in watchmaking. Below, a model in white and yellow gold. Opposite, a square model for men and a round model for ladies. All have the 7P movement.

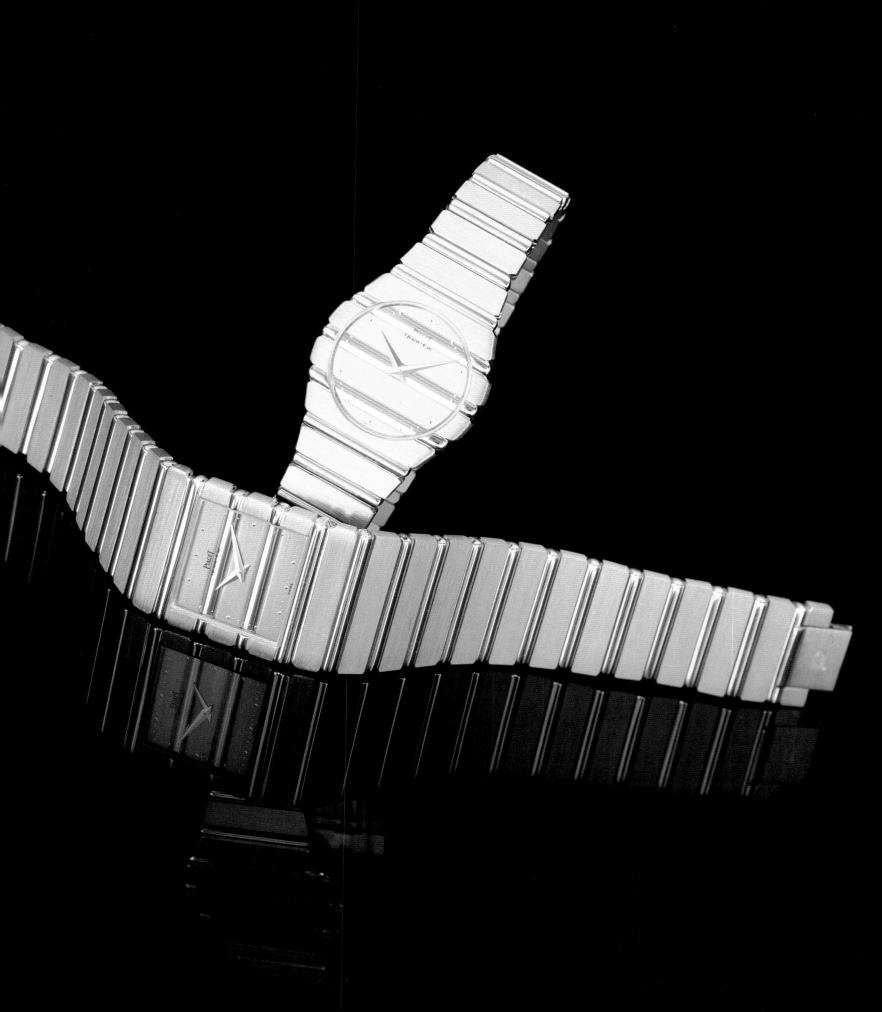

This automatic model with date display has a yellow gold case and bracelet (total weight 96 grams). 12PC movement (1974).

Another automatic model, this one without date. The tiger-eye dial features gold Roman numerals. 12P movement (1979).

Gold bracelet watch for ladies with an onyx dial set with twelve diamonds that function as hour markings. 9P movement (about 1973).

Ultra-thin ladies' model with bezel and bracelet engraved with the same design. Note the small cabochon on the crown. 9P movement (1974).

The rectangular dial, in malachite, of this large bracelet watch is highlighted by two plaques of the same gemstone. 9P movement (1977).

Left: An elegant evening model for men in white and yellow gold with onyx dial. Bezel and bracelet are set with diamonds. 7P quartz movement (1972).

Below: This exceptional high-jewelry watch has gold case and bracelet set with diamonds and baguette-cut sapphires 4P movement (1979).

Thus end the 1970s, the decade that saw an explosion of ready-to-wear and new philosophies, the legendary gatherings of young people such as Woodstock, the psychedelic subculture, and the feminist wars, Giorgio Armani on the cover of *Time* Magazine and the Japanese the winners in the electronic wars in every field, from multifunction watch to the Walkman. The economic crisis, already on its way to resolution in the wealthy countries, turned into a crisis of conscience. A certain rigor leveled the field of objects: Italian design prevailed the world over, while certain products of French technology, like the supersonic Concorde and the supertrain TGV, placed it in the forefront. The mood was to take advantage of life, to spend on beautiful objects, especially those that lent a certain image, the key word of the 1980s. In its area, Piaget found itself in a strong position to take up the new challenge.

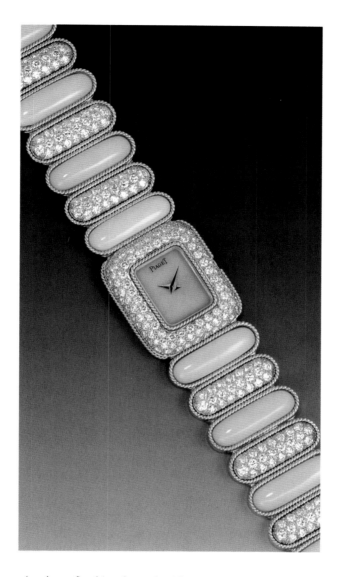

Another refined jewel-watch with case and bracelet in white and yellow gold, as well as 303 diamonds and 10 "angel-skin" cabochon corals. 4P movement (1978).

This model with case and bracelet in white gold is embellished in the center of the dial with pavé-set diamonds, while 28 additional diamonds decorate the oval bezel. 9P movement (1979).

One of three models created in a limited and numbered edition to celebrate Piaget's hundredth anniversary. Each of these rare watches is "crowned" by a 1-carat diamond. This example, marked number 4, is distinguished by its onyx dial with diamonds used for hour markings. 9P movement (1974).

Opposite: Yves G. Piaget and Gina Lollobrigida, the famous Italian actress (and admirer of Piaget watches), are engaged in conversation with the sculptor César.

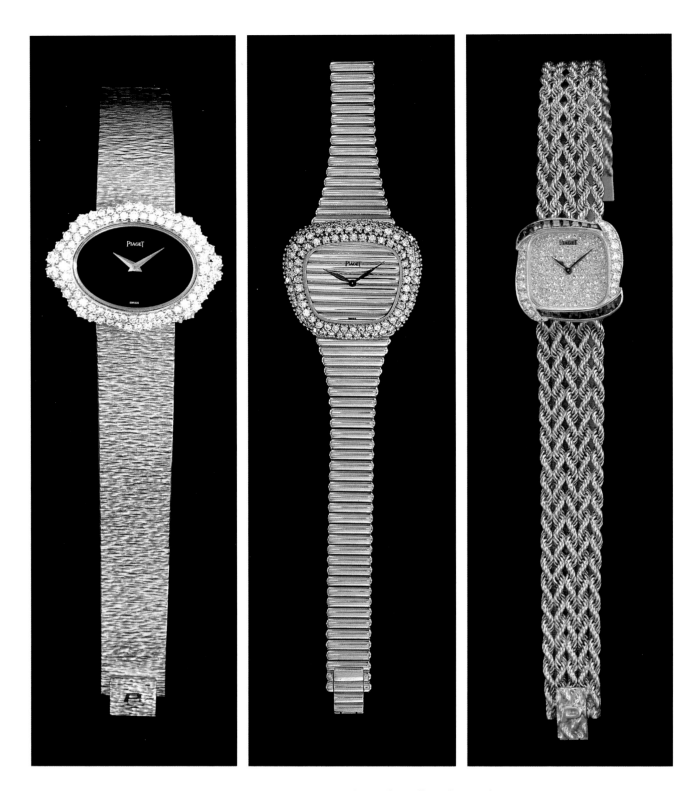

Three interpretations on the theme of "small jewelry watches for ladies" in gold. Left, an onyx dial surrounded by a diamond "petticoat" carries the 9P movement; center, the harmonious curve of lines for a bracelet that also hides the 9P movement; right, a delicate shape sets off the bezel set with baguette-cut emeralds and diamonds, 4P movement (1974, 1976, and 1979).

The "river" shape of the bracelet flows around the "islet" of the oval case with lapis lazuli dial and prominent bezel decorated with diamonds. 4P movement (1977).

Above: No fewer than 2,125 diamonds were used to cover the case and bracelet of this model in white gold. The dial displays another 156 diamonds and 8 sapphires. Beta 21 quartz movement (1972). Below: Another jewel-watch in white gold. 7P quartz movement (about 1976).

On this page: This exceptional interpretation on the theme of the high-jewelry watch in white gold is set all over with diamonds edged with baguette-cut emeralds. 7P quartz movement. The watch, created in 1978, cost 461,000 Swiss francs at the time.

Opposite: It often comes about that the creation of a watch leads to a complete parure. This ensemble of watch, necklace, ring, and earrings offers an admirable example of Piaget's mastery of the art of mounting and assembling precious stones; here, the stones are diamonds and cabochon emeralds. The watch dial is mother-of-pearl. 9P movement (1980).

Above left: This octagonal watch with satin-finish gold case has a steel blue dial with applied gold hour markings. 9P movement (1973). Above right: Another octagonal case, this one wider, with bezel and bracelet chased in the same design and lapis lazuli dial. 9P movement (1973).

Left: Bracelet watch in yellow gold, with bezel and bracelet of the same design. The center of the onyx dial is set with diamonds. 9P movement (1975).

Above, left to right: Bracelet watch in white gold with blue tiger-eye dial, the center of the dial set with diamonds, 12P automatic movement (1975); rectangular model with onyx dial, diamonds functioning as hour markings, and a center with pavé set diamonds, 9P movement (1978); another rectangular model with leather strap, also with onyx dial and carrying four diamonds as hour markings and a bezel set with diamonds, 9P movement (1973).

Below left: This model with case and bracelet in yellow gold has a dial entirely set with diamonds and framed in onyx. 9P movement (1976). Below right: This model with hexagonal case and bracelet in white gold has a ruby dial with a diamond set center. 9P movement (1975).

Pressing lightly on the button on the side reveals the watch hidden inside the bracelet. The dial is in tiger-eye with gold Roman numerals. This bracelet watch, made in a single model and weighing 200 grams, has a maximum width of 34.2 mm and a maximum thickness of 10.3 mm. It carries a 9P-caliber movement (1972).

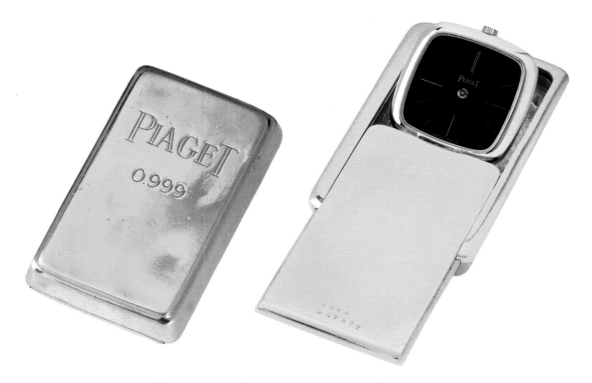

By sliding the cover, this gold ingot turns into a refined pocket watch – another facet of the Piaget magic. The ingot is in 24-karat gold, the case of the watch in 18-karat gold; black dial. 9P movement (1974).

The dial and six medallions that compose the bracelet of this watch present the same pre-Columbian motif – a homage to the ancient civilizations of South and Central America. The bezel carries a sun-ray pattern. 9P movement (1971).

Above left: Case and bracelet in white gold, bezel and onyx dial embellished with diamonds, the center part of the dial studded with diamonds. 12P automatic movement (about 1974).

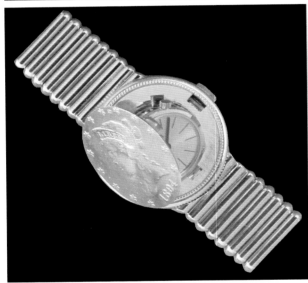

Center left: A very rare coin watch for ladies meant to be worn on the wrist, with gold bracelet as well. The coin is a ten-dollar piece from 1894; the dial is gold. 4P movement (1977).

Below left: This bracelet watch has case, dial, and bracelet engraved in the same motif, and it is in Mexico called "Emperor". The square watch whose case has the same width as the bracelet is highly characteristic of Piaget models. 9P movement (1972).

Above right: The case of this gold watch features rounded corners and a bezel that is both polished and chased. The jade dial carries gold minute dots. 7P quartz movement (1977).

Below left: A refined ring watch in white gold with stepped bezel set with 395 round diamonds, whose brilliant reflections set off the black dial. 4P movement (1977).

Below right: Case and bracelet in white gold, with lapis lazuli inlaid in the links and on the bezel alternating with diamonds. The dial is also in lapis lazuli and uses four diamonds as hour symbols. 7P quartz movement (1977).

Opposite: Another elegant gold watch, whose bracelet in "bars" prefigures those of the "Polo" line. The onyx dial has four diamonds at the cardinal points. 7P quartz movement (1972).

A highly refined model for ladies in white gold. The oval dial, made of lapis lazuli, is framed by a double row of diamonds set in a herringbone pattern. 7P quartz movement (1976).

Another jewel-watch with lapis lazuli dial. The diamonds that frame the dial extend along the bracelet. Unlike the watch at left, this one has a mechanical movement, the 4P caliber (1977).

Around the lapis lazuli dial, a brilliant display of the jeweler's craft: a double row of marquise-cut diamonds that extend beyond the dial to overlap the bracelet. Mechanical 9P movement (1978).

Following page: A creation by one of a group of "avant-garde couturier-designers" who showed the most extravagant fashions in New York, combining rags, plastic bags, old postcards, even hair curlers – the newest fashions of surrealist haute couture.

Quartz: Absolute Precision

The advent of quartz instigated a true revolution in the watchmaking industry, comparable to the introduction of the pendulum around 1670. The latter permitted the making of clocks no more than ten seconds off a day; in the twentieth century, the quartz narrowed the margin of error to a previously unimaginable thousandth of a second a day. However, the analog quartz movements – those that utilize watch hands – are largely similar to traditional mechanical movements, with wheels, pinions, and rackets. What changed completely was the method of measuring time. In a mechanical watch, that task falls to the vibrating unit, that is, the balance, which might vibrate 14,400 to 36,000 times an hour, or accomplish 2 to 5 oscillations a second; the more constant its "beat," the more precise the watch will be. However, in many cases, especially in bracelet watches, the regularity of the balance's vibrations can be disturbed by external factors, as, for example, the numerous changes in position to which we submit our watches by moving our arms; abrupt changes in temperature or atmospheric pres-

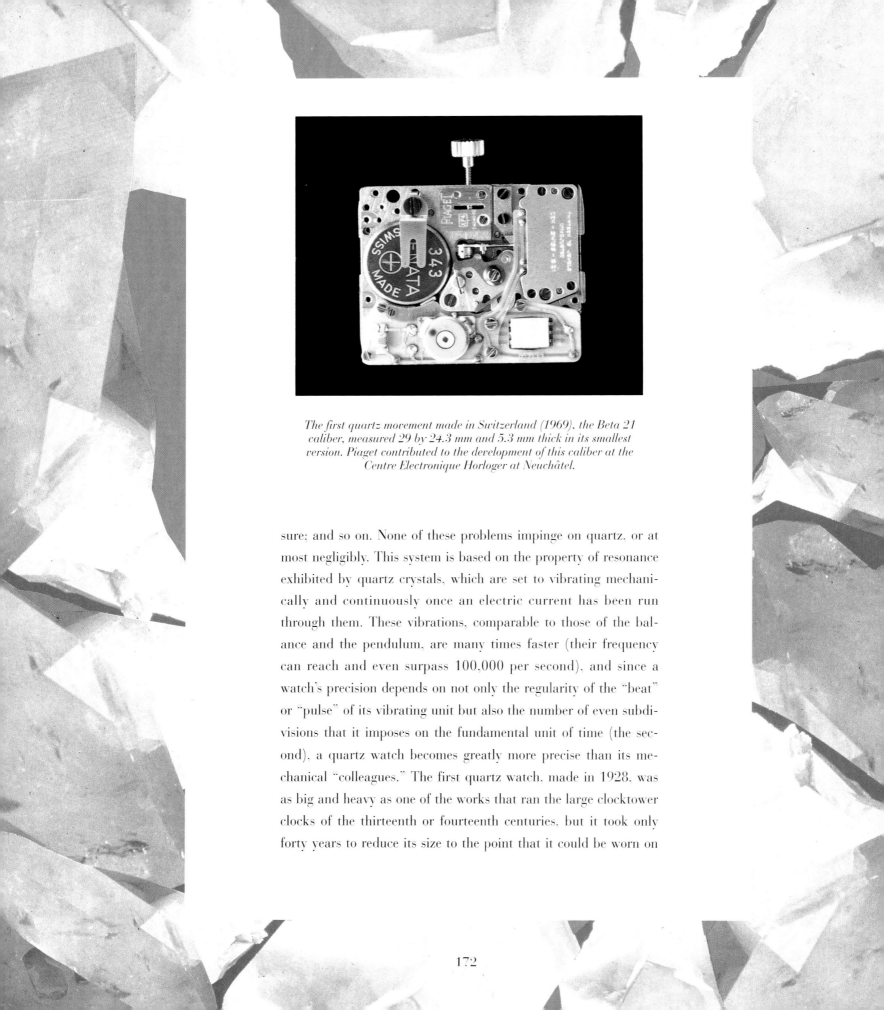

The first quartz movement made in Switzerland (1969), the Beta 21 caliber, measured 29 by 24.3 mm and 5.3 mm thick in its smallest version. Piaget contributed to the development of this caliber at the Centre Electronique Horloger at Neuchâtel.

sure; and so on. None of these problems impinge on quartz, or at most negligibly. This system is based on the property of resonance exhibited by quartz crystals, which are set to vibrating mechanically and continuously once an electric current has been run through them. These vibrations, comparable to those of the balance and the pendulum, are many times faster (their frequency can reach and even surpass 100,000 per second), and since a watch's precision depends on not only the regularity of the "beat" or "pulse" of its vibrating unit but also the number of even subdivisions that it imposes on the fundamental unit of time (the second), a quartz watch becomes greatly more precise than its mechanical "colleagues." The first quartz watch, made in 1928, was as big and heavy as one of the works that ran the large clocktower clocks of the thirteenth or fourteenth centuries, but it took only forty years to reduce its size to the point that it could be worn on the wrist. The Swiss industry was very aware of the challenge posed by the new technology when it decided to take it up: in 1962 it created CEH, the Centre Electronique Horloger, at Neuchâtel. Seven years later, Piaget, one of the prestigious companies that contributed money and participated in the research and development devoted to "electronic time," unveiled its first quartz models. These were two watches for men of large dimensions, as their movement, the Beta 21 caliber, was considerably thicker than a mechanical movement like the 9P or the 12P. On the other hand, it was much more precise: the quartz crystal, placed in a vacuum-packed capsule and given an elastic mount to protect it from shocks, vibrated 8,192 times a second; an integrated circuit, or chip, would channel these vibrations and bring them down to 256 a second; then a micromotor of vibrations clicked into action, transmitting the impulses by means of a ratchet bar to a wheel with 256 teeth that made a complete revolution in one second; a

Piaget immediately began working toward miniaturizing the quartz movement. In 1976 it came out with the 7P caliber, four times smaller than the Beta 21.

train of gears reduced the rapid rotation of this wheel and sent it to the hands, then, from these, to the calendar disk. A battery of 1.35 volts, which lasted about fourteen months, kept it going, but the real "brain" of the watch resided in the integrated circuits, which, small as they were, contained enough transistors to run five radios. Tests made at the Neuchâtel Observatory confirmed that the daily loss of the Beta 21 measured in tenths of seconds, or less than a minute a year. The great precision of quartz instruments rendered the yearly contests for the most accurate chronometer academic; henceforth, a greater than remarkable precision had been harnessed, more than sufficient for the ordinary, nonscientific use that a bracelet watch called for. But, true to its motto, "Always do better than necessary," Piaget carried out at La Côte-aux-Fées a series of studies and projects aimed at improving the new technology in its turn. These years of investment and research paid off in 1976, when Piaget presented the 7P caliber, the thinnest quartz movement in the world. In a body 3.1 millimeters thick, the new

In 1976, Piaget offered the thinnest quartz watches in the world, equipped with the 7P caliber. Shown life-size above is a scaled-down version of the model 14101, which had been designed in 1969 for the Beta 21 caliber.

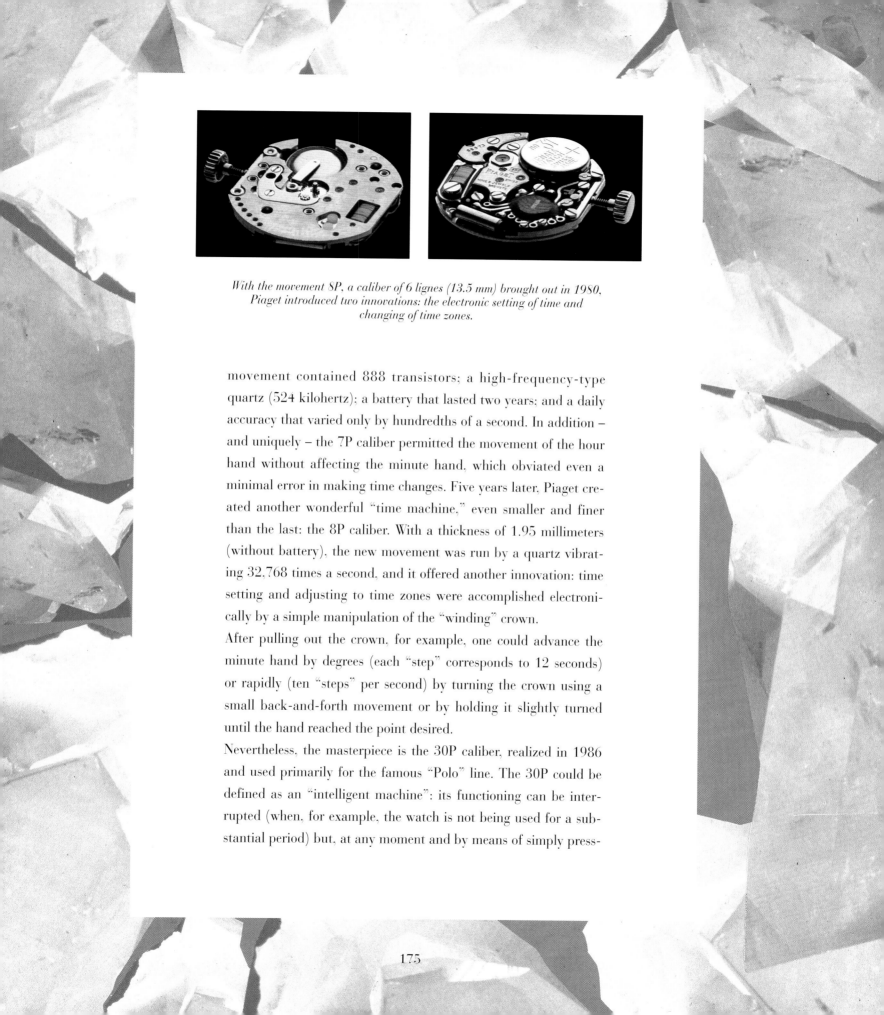

With the movement SP, a caliber of 6 lignes (13.5 mm) brought out in 1980, Piaget introduced two innovations: the electronic setting of time and changing of time zones.

movement contained 888 transistors; a high-frequency-type quartz (524 kilohertz); a battery that lasted two years; and a daily accuracy that varied only by hundredths of a second. In addition – and uniquely – the 7P caliber permitted the movement of the hour hand without affecting the minute hand, which obviated even a minimal error in making time changes. Five years later, Piaget created another wonderful "time machine," even smaller and finer than the last: the 8P caliber. With a thickness of 1.95 millimeters (without battery), the new movement was run by a quartz vibrating 32,768 times a second, and it offered another innovation: time setting and adjusting to time zones were accomplished electronically by a simple manipulation of the "winding" crown.

After pulling out the crown, for example, one could advance the minute hand by degrees (each "step" corresponds to 12 seconds) or rapidly (ten "steps" per second) by turning the crown using a small back-and-forth movement or by holding it slightly turned until the hand reached the point desired.

Nevertheless, the masterpiece is the 30P caliber, realized in 1986 and used primarily for the famous "Polo" line. The 30P could be defined as an "intelligent machine": its functioning can be interrupted (when, for example, the watch is not being used for a substantial period) but, at any moment and by means of simply pressing the button on the back of the case, the hands find the exact hour, as if by magic. In addition, the exact date appears in the tiny window of the date display, even if the watch has not been used for several months.

In effect, the 30P caliber is also a perpetual calendar – that is, its mechanism automatically keeps track of months with 30 or 31 days as well as leap years. Technically, this is accomplished by a "complication," that is, an additional mechanism, up to that point reserved only for some mechanical models. But Piaget, using a tiny integrated circuit with 24,000 transistors and endowed with "memory," succeeded in joining it to the extraordinary precision of quartz, adding another improvement to its elegant models.

Quartz technology has continued to evolve: in October 1992, an even more complex and complete quartz analog movement saw light of day: the 212P caliber, the smallest chronograph with flyback function and perpetual calendar. Known under the code name "medico" during its development (it seems that it was a doc-

This circuit, shown here magnified many times (it actually measures 4.5 by 4.5 mm), with 24,000 transistors, drives the 30P caliber. Introduced in 1986, it is the world's only disk-type perpetual calendar watch.

watches that once again created a sensation. In 1979, it had introduced "Polo," water-resistant and presenting an extra-thin case, worked in 130 grams of solid gold, and employing the brand-new and prestigious 7P quartz movement, the only one in the world to possess a memory of the exact time and instant changing of time zones. And in 1981 Piaget's success as jeweler was appropriately celebrated with the "Phoebus," the most expensive watch in the world (3.5 million Swiss francs at the time), obviously made in one model only, ordered by a Japanese. The fruit of two years of work on the part of Piaget's master jewelers in Geneva, it made its premiere in a Red Cross gala in Monte Carlo, in the presence of Prince Rainier and Princess Grace of Monaco, with Sylvia Krystel and Yves G. Piaget in attendance. With 154 grams of platinum, studded with 296 exceptionally pure diamonds (for a total of 87.87 carats), including a blue diamond of 3.85 carats whose luminosity made it unique, the watch clearly earned its name, which derives from Apollo, the sun god of Greek mythology. This fortunate collector was not the only one to choose Piaget. Many from the world of princes, sheiks, and Arabian emirs, who have demonstrated a strong attachment to exceptional products such as the jewel-watch, personally commissioned Yves G. Piaget or his closest colleagues for wristwatches or pocket watches carrying the family arms in diamonds, rubies, and emeralds.

One prince ordered the same type of watch in the version for ladies for each of his daughters, who at the time numbered twelve. For the heir to the throne of Bahrain, two splendid bracelet watches were created, with the country's arms engraved on the dial in gold, mother-of-pearl, and coral. For an Arab client, at his order and using his own design, Piaget struck medals in trapezoid form to commemorate the 1,400 years since the Hegira, the flight of the prophet

Above: Even the rotor of the automatic winding mechanism has been finely engraved on this yellow gold model with perpetual calendar and moon phases. Below: the same watch seen from the dial side.

Opposite: Another skeleton model, this one a pocket watch, with a partially cutout dial that reveals the movement and the decoration of the bridges. Note the graceful form of the pendant and the cabochon crown (1983).

Above: Skeleton bracelet watch with extra-thin case. The bezel, set with 34 diamonds, is framed by rock crystal. 9P movement (1985).

Below: Another extra-thin skeleton watch with a special attachment to the bracelet. The 20P movement is 1.2 mm thick and is surrounded by a circle of pavé-set diamonds (1982).

Opposite: Skeleton version of a chronograph. A refined model, it has a hobnail-pattern on the bezel and small push-buttons. Mechanical movement, 27mm in diameter (1987).

On this page: This high-jewelry watch carries 175 sapphires (a total of 28.53 carats) on its bracelet, 22 diamonds on its bezel, and a mother-of-pearl dial (1985).

Opposite: Another splendid parure, consisting of watch, necklace, earrings, and ring, made in yellow and white gold. Its decorative effect was created by setting the entire surface with diamonds and emeralds (about 1982).

 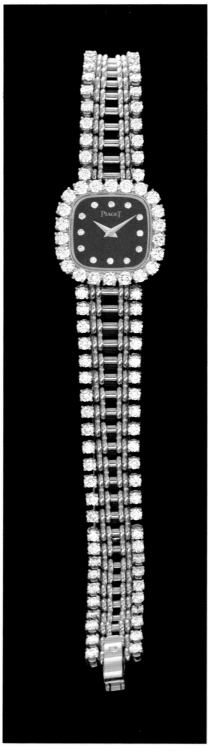 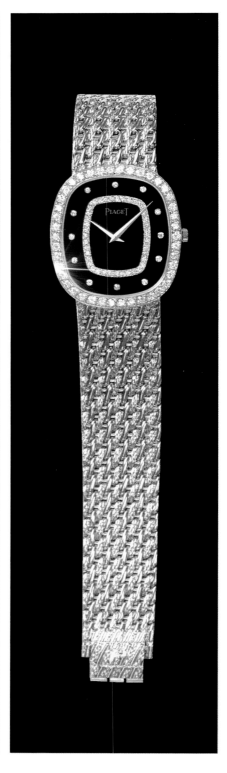

Watch and bracelet in yellow gold with central links and bezel set with diamonds. Mother-of-pearl dial with diamonds and rubies. SP quartz movement (about 1984).

Onyx and diamonds for this jewel-watch in yellow gold with unusual bracelet. On the onyx dial, 12 diamonds serve as hour symbols. SP quartz movement (1985).

Again, diamonds (42 in all) and onyx for this model with oval case and yellow gold bracelet. The 9P-caliber movement is mechanical (about 1983).

Unusual model with oval dial and bezel formed by a triple band of diamonds in the shape of a ribbon. The dial is set with diamonds as well. SP quartz movement (1984).

Muhammad to Medina in 622 A.D. Numbered from 1 through 1,400 and mounted mostly in pendants, some of them could not fail to make their way into bracelet watches.

The very close and guarded circle of oil kings, which came to dominate world economies once again at the beginning of the 1980s (with the Gulf War a distant and unforeseeable event), constituted one of Piaget's most important groups of clients. A very different group, personalities very much in the public eye, chose Piaget since the end of the 1970s for the joy of owning truly original and unique models. The pianist Liberace showed off his watch with bracelet in the form of a grand piano, the keyboard composed of baguette sapphires and diamonds. Andy Warhol had a remarkable collection of Piaget watches. The watch made for writer Frédéric Dard, author of the San Antonio series (the name is inscribed on the upper edge of the case), took the form of a book, with a black dial inscribed "Minuit" (midnight). Adnan Khashoggi, one of the richest men in the world, known to magazine readers for his love affairs and his boats, has been a faithful client of Piaget for twenty years. However, one of the oddest and most unexpected owners of a Piaget watch is Fidel Castro, who has the model that Gina Lollobrigida wore – that is, a ladies' model. The Cuban leader had invited her to visit his country. Noticing her Piaget, he proposed an exchange: his old watch, which had gone through many battles with him, for hers. Only time will tell who got the better deal – if history is worth more than diamonds. In addition, the magazine *Town and Country* revealed that at the beginning of the 1980s the most beautiful society ladies of Florida, from the white houses of Palm Beach to the villas of Coconut Grove, preferred to wear a Piaget with a dial of semi-precious stone – coral, naturally, which shows up well against the peach-colored tan skin of Americans. In

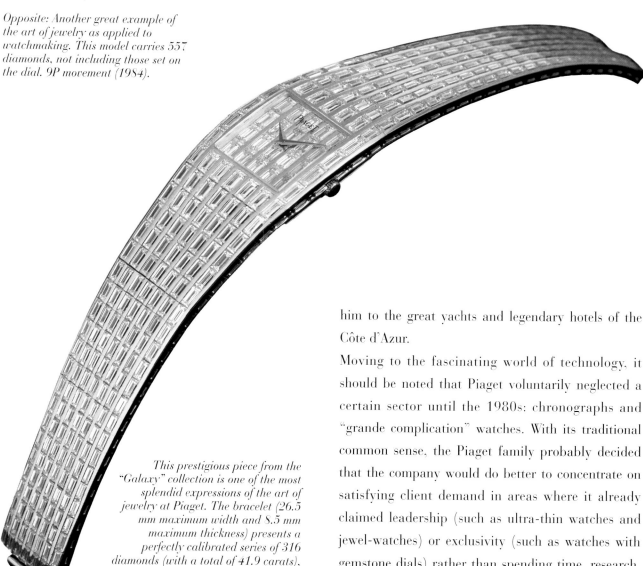

Opposite: Another great example of the art of jewelry as applied to watchmaking. This model carries 557 diamonds, not including those set on the dial. 9P movement (1984).

This prestigious piece from the "Galaxy" collection is one of the most splendid expressions of the art of jewelry at Piaget. The bracelet (26.5 mm maximum width and 8.5 mm maximum thickness) presents a perfectly calibrated series of 316 diamonds (with a total of 41.9 carats), and the dial is covered with 20 more. 9P mechanical movement (1987).

1976. Yves G. Piaget had selected Palm Beach, which is to the latest generation of Kennedys as Martha's Vineyard was to John and Jackie, for his Polo Championship. This brought together the best teams in the world and offered prize money of $100,000. Yves spent the winter near some of his best clients at Gstaad and Saint-Moritz, while the summer brought him to the great yachts and legendary hotels of the Côte d'Azur.

Moving to the fascinating world of technology, it should be noted that Piaget voluntarily neglected a certain sector until the 1980s: chronographs and "grande complication" watches. With its traditional common sense, the Piaget family probably decided that the company would do better to concentrate on satisfying client demand in areas where it already claimed leadership (such as ultra-thin watches and jewel-watches) or exclusivity (such as watches with gemstone dials) rather than spending time, research, and capital trying to challenge the hegemony of other important companies already famous for their "grande complication" watches. However, with the boom in luxury watches that became so pronounced in the 1980s, it seemed wise not to neglect any area. Thus appeared at opportune moments several splendid models, like the skeleton chronograph, the calendar watch with complete moon phases, realized in a limited and numbered edition (250), or the beautiful "complication" model with date, day, month, and phases of the moon, white dial and gold bracelet (only ten examples produced). In addition, there were several precious pocket watches, with gold case, dia-

This model with bracelet is distinguished by the unusual outlining of the case and by its dial in mother-of-pearl inlaid with gold and set with diamonds. The bezel and the links of the bracelet also sparkle with diamonds.
SP quartz movement (1982).

monds, and chronometer-type precision movement, as well as "skeletons" of unusual beauty.

In 1982 and 1983, Piaget once more made records in the realm of the ultra-thin. After numerous months of studies and research, it developed a mechanical movement – the 20P – with a diameter of 20.4 millimeters and a thickness of 1.2 millimeters. The Guinness Book of World Records named it the thinnest mechanical movement in the world. This "one-piece" (*monobloc*) caliber has a single plate in which the moving pieces are mounted on one side only in miniature ball bearings.

Using the 20P as a base, the automatic caliber 25P was built, which superimposed a second plate that carried the rotor (oscillating weight) and the automatic winding mechanism. Its thickness of 2 millimeters made it the thinnest in the world, having shaved off three-tenths of a millimeter since Piaget's last record of 2.3 millimeters, won by the 12P caliber in 1960. Despite these purely horological feats, Piaget did not take these two calibers into full production, as they did not meet the traditional strict criteria of quality that Piaget demanded before considering their commercialization.

One might be tempted to ask, from a practical point of view, what purpose such performances serve, since the difference can barely be perceived by the naked eye. The answer holds equally for other technological feats as well as in sports: the passion to set records, to cross the limits of what is possible, to rise to the challenge of not only the contest but also the laws of the mechanics of precision.

When it comes to sales, the setting of records contributes to the image of high-tech watches. On the other hand, the ultra-thin, as we have seen (in the 9P, 12P, and the quartz 7P movements), allowed the designers and jewelers of Geneva (whose numbers, in the 1980s, due to a series of recruitments during the

The alternation of textured and polished gold sets apart this bracelet watch with "checkerboard" dial. SP quartz movement (1985).

Another refined marquetry effect gained by the alternation of textured and polished surfaces. This watch also carries the famous SP quartz movement (1985).

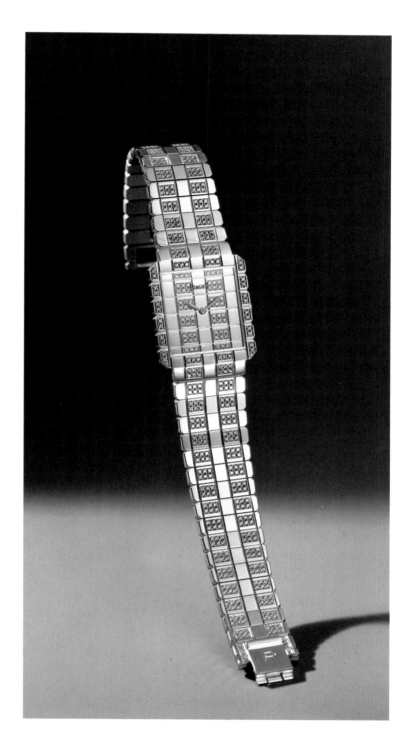

Case, dial, and bracelet carry the same design in this elegant evening model, which has diamonds on its dial and on the bracelet as well as the sides of the case. 9P mechanical movement (1985).

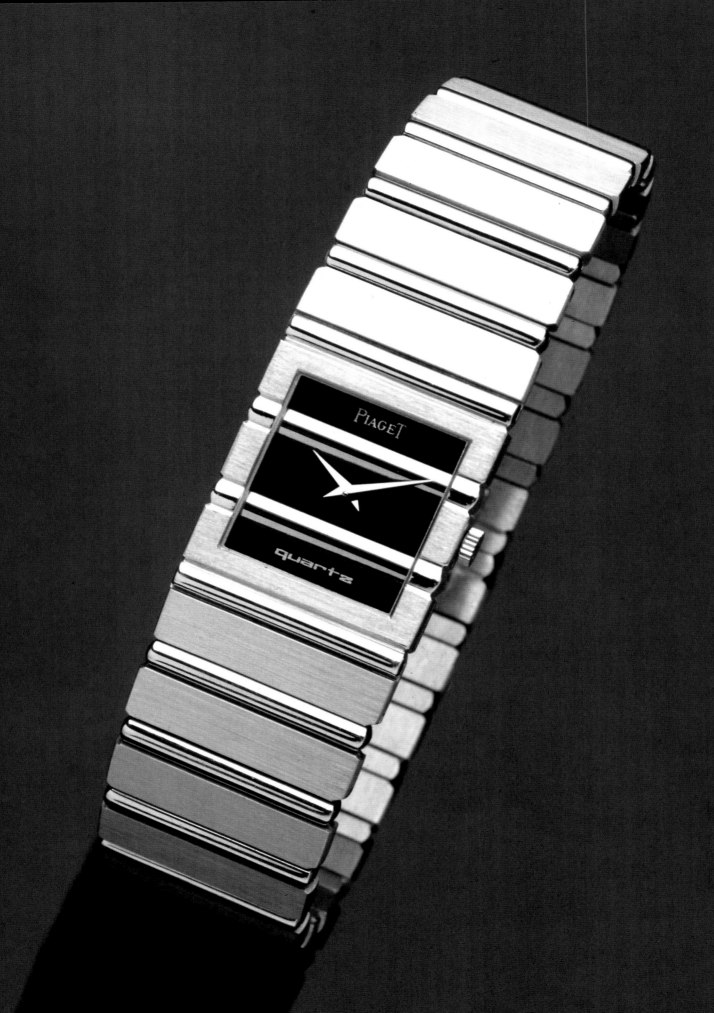

This "Polo" model (the smallest of all those with a round case) has diamonds on the bezel and on the onyx dial. Note that the attachment of the bracelet is invisible. 4P movement (1985).

Another "Polo" model, this one in white gold with a square case. Bezel and dial are entirely set with diamonds. The attachment of the strap is invisible. 9P movement (1985).

Satin-finished gold and three bands of diamonds embellish the dial and the case of this other version of "Polo" for ladies. SP quartz movement (1985).

The largest of the "Polo" models with round case. This one is characterized by the contrast between the satin-finished and polished gold of the case and dial. SP quartz movement (1985).

Opposite: A square-case "Polo" in white gold, in one of its simplest and most refined variations. Two of the bracelet's polished ribs cut across its onyx dial. Quartz caliber SP (about 1985).

The incomparable "Polo" in its square and round versions. The watch on the left is in yellow and white gold, decorated with diamonds. 8P quartz movement. The watch on the right is in yellow gold and alternates 158 baguette-cut diamonds and 500 round diamonds. 7P quartz movement (1984). The cuff links to its left mirror the style of the watch.

1960s and 1970s, exceeded those of master watchmakers at La Côte-aux-Fées) to realize watches that, from the aesthetic point of view, were equally spectacular.

The most beautiful example of these jewel-watches, worthy of a place in a museum of decorative arts or a royal collection, is given by the complete collection called "Ermitage," with twenty different pieces. All take the form of skeleton-type pocket watches, based on the perfect shape of the circle. The movement, however, often is placed in original shapes, notably, a pear limned by a lace of diamonds and emeralds, or a small perfume flask in rock crystal, diamonds, and rubies. The name of the collection indicates a homage to art, referring to one of the most remarkable museums in the world, and perhaps also to the Russia of the czars and of Fabergé, by means of these watches that evoke an enchantment comparable to that wrought by the famous Easter eggs in enamel and precious stones. In addition, as Yves G. Piaget explains, "The Ermitage collection was not conceived with an eye to sales but to our own Piaget museum and traveling exhibitions that we were planning to

Below: Square model in yellow gold with white dial whose gold Roman numerals are applied. SP quartz movement (1985). Next to it, a watch in the classic round shape whose bezel also displays the same finish as the bracelet. SP quartz movement (1984).

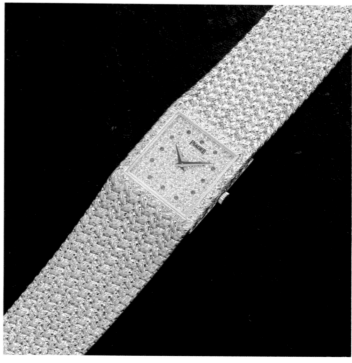

Above: Two versions of jewel-watches with rubies as hour markings. The bezel of the oval model is set with 40 diamonds. Both carry the 9P mechanical movement (1985).

This very original bracelet watch decorated with 99 diamonds has a heart-shaped gold case. The bracelet consists of pearls and rubies (a total of 44). 8P quartz movement (1987).

organize in different cities throughout the world in homage to the excellence of our watchmakers and jewelers. Of course, if one of our clients wanted to have one of these pieces immediately, we would have satisfied that wish..." When Yves G. Piaget gave that interview in *Town and Country*, the year was 1983, and the prices ranged from $100,000 to $3 million.

In 1984, Ronald Reagan was reelected to the White House. At Lucerne, the artist Hans Erni paid homage to the Piaget family history by creating the "Piaget gold piece," a "coin" of four sizes and denominations, from the small demi-Piaget gold piece to the five-Piaget gold piece. The figures on the obverse and reverse of the coins were taken from classical symbolism. Struck in pure gold by the Federal Mint in Bern in a limited and numbered edition, these pieces of gold were mounted on pendants, bracelets, cuff links, and so on, or sold to collectors; only a few examples of the two- and the five-Piaget gold piece were transformed into watches.

Naturally, Piaget devotes just as much attention to its most popular models as its unique, museum-quality masterpieces. After all, the popular watches form the backbone of the daily sales figures. Among this category, the model "Polo" has long performed as the company's best-seller, due as much to its image as to the way Piaget has succeeded in updating it, presenting it each year with a new series of models. Introduced in 1979, in the 1980s it enjoyed a phenomenal success, from the classic gold "Polo" with a gold bracelet to those given a gemstone dial; from the version with a leather strap designed with vertical rather than horizontal bars to the high-end jeweled models, such as the ones embellished with diamonds. In 1986, a high-tech model was brought out, an ultra-thin perpetual calendar (with Piaget's 30P quartz movement): by pressing a small button hidden on the back of the case, one can stop the movement if the

This piece of watch jewelry boasts gold, diamonds, rubies, pearls, and emeralds in the shape of pearls. SP quartz movement (1984).

watch will not be used for a while. Even if months or years pass before putting it back on the wrist, pressing the button once more starts it up again, and it automatically resets itself to the correct date and time. In addition to these practical features, the perpetual calendar "Polo" (which, of course, comes in a version for ladies as well) keeps track of leap years.

An entire new family of watches, the "Dancer" line, made its appearance in 1986. At a certain point in its development, it seemed destined to exceed the sales figures of "Polo." In its extremely classic design, very characteristic of Piaget, especially in the basic models, it favors the round case, which has seen a renewed interest. In its special collections, especially the models in gold with gold and diamond bracelet, the square "Dancers" are much appreciated for the harmony of their lines. As usual, a great variety of dials are available, with mother-of-pearl and other stones coming into play. Most versions can be outfitted with either mechanical or quartz movement.

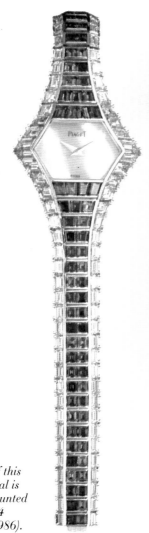

The hexagonal-shaped case of this watch with mother-of-pearl dial is subtly framed by the finely mounted setting that accommodates 104 diamonds and 98 emeralds (1986).

Another virtuoso interpretation on the theme of the heart-shaped watch, this one-of-a-kind piece was made with 101 diamonds and 120 rubies. 4P mechanical movement (1987).

A skeleton pocket watch with 40 diamonds and 8 sapphires in a notched circle beneath the crystal, held in place by a beaded bezel (1981). Below, left: an automatic model with perpetual calendar that indicates the leap year and the phases of the moon (1982). Right: Another automatic watch for men, with classic date display and sweep seconds (1982).

This ultra-thin model is only 3 mm thick. It has a beaded bezel and a 20P caliber mechanical movement – the thinnest in the world, at 1.2 mm. Gold dial with painted hour markings (1982).

Another ultra-thin model, this one carrying one of the thinnest automatic movements in existence (2 mm): the 25P caliber. Slim, stepped bezel with classic white dial (1982).

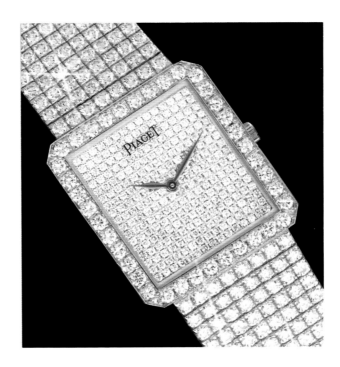

This exceptional time-piece is entirely set with diamonds; in addition to those that decorate the dial, 577 diamonds cover the case and the bracelet. 9P movement (about 1985).

Opposite: An engine-turned "checkerboard" design on the bracelet, bezel, and dial characterizes this refined yellow gold bracelet watch. 9P movement (1986).

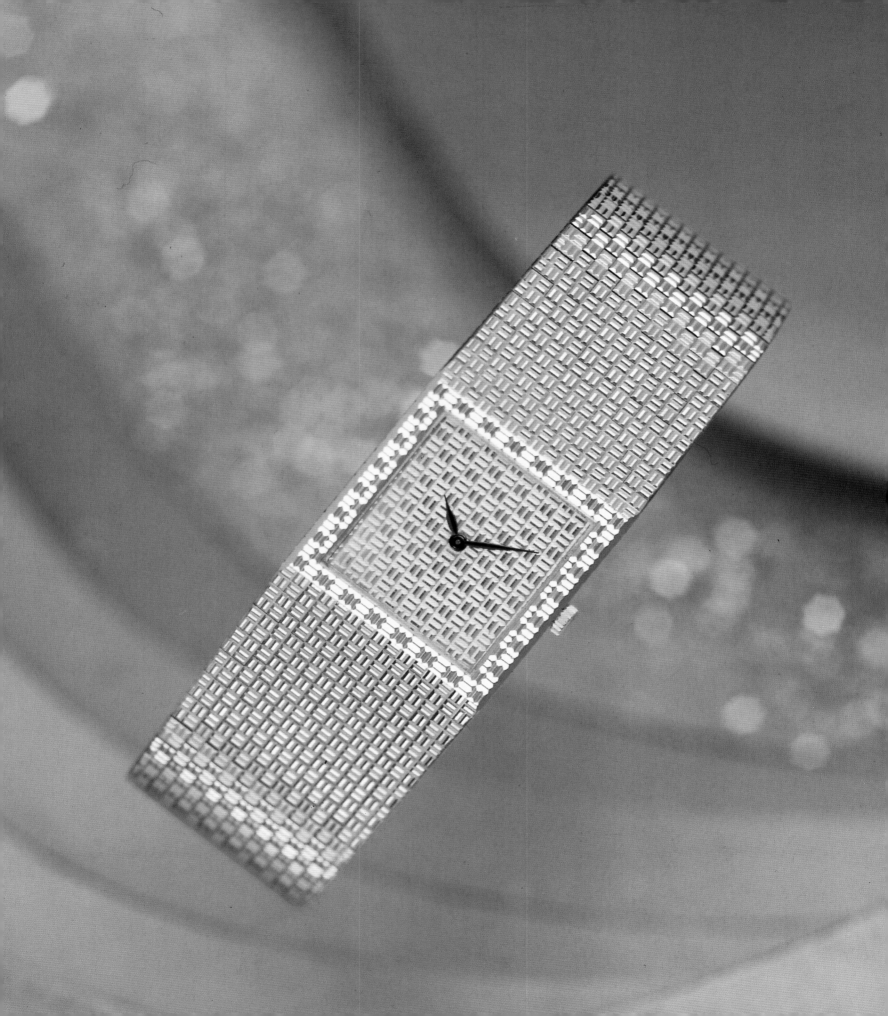

Left: The central links of this jewel-watch are set with diamonds. The mother-of-pearl dial has four diamonds as hour markings. 601P quartz movement (1985).

Right: Diamonds embellish the bracelet and bezel of this model with oval case. The dial is set with diamonds and 4 ruby hour markings. 4P mechanical or 8P quartz movement (1985).

Left: Five rows of gold links give the bracelet its design. In addition, dial, bezel, and bracelet are set with diamonds. 4P mechanical or 8P quartz movement (1986).

Right: Another jewel-watch for ladies, its gold case embellished with diamonds and its dial having the same design as the bracelet. 4P mechanical or 8P quartz movement (1986).

The bracelet with large convex links, a motif repeated on the case, distinguishes these three square models. All carry quartz movements (about 1984).

Yves G. Piaget and Emil Keller converse with the painter Hans Erni.

President Ronald Reagan warmly greets Yves G. Piaget.

Camille Pilet checks the finish on two newly completed bracelet watches.

At the Monte Carlo tournament in 1973, Yves G. Piaget presents the famous tennis player Mats Wilander with his prize, a prestigious watch, as Marie-France Pisier looks on.

The play of white and yellow gold characterizes the bracelet and bezel of this "Dancer" model, which has a mother-of-pearl dial with 12 diamonds serving as hour markings. SP quartz movement (1985).
Right: A different version of "Dancer," this one square, whose dial and case continue the cylindrical motif of the bracelet. SP quartz movement (1986).

Another version of "Dancer," always with cylindrical motif, but here carried out in white gold. The outer rim of the dial is in onyx and uses 12 diamond hour markings; the center of the dial is studded with diamonds. More diamonds decorate the links and the bezel. 9P movement (1986).

Opposite: This yellow gold "Dancer" has a round dial and bezel embellished with diamonds. Its bracelet displays the cylindrical motif typical of the watches of this collection. SP quartz movement (1986).

Above: A selection of "Dancer" models with leather strap and classic round case in yellow gold, with contoured lugs and bezel engraved with the twelve hour symbols typical of the "Dancer" family. The dials are greatly varied, also evident in the "Dancer" models of the previous pages. The watches above carry the mechanical 9P movement or the quartz 8P (1986).

Below: At left, a classic round case in yellow gold and white dial with small gold Roman numerals. 8P quartz movement (1986). At right, a refined bracelet watch, with perpetual calendar date display. 30P quartz movement (1987).

Above: Gold case and dial, with perfectly smooth bezel. On the outer rim of the dial, four pairs of diamonds mark the cardinal points. 9P movement (1987).

Below: At left, an ellipse-shaped watch with yellow gold case and slender bezel. The gold hours are applied on the gold dial. 9P movement (1985). At right, the hexagonal case has rounded corners. The white dial carries painted Roman numerals. 9P movement (1982).

Left: The watch on the left was produced in a limited and numbered edition of 250. It has complete calendar (day, month, and date), moon phases, and seconds hand. 27P mechanical movement (1985). The watch on the right is the prototype of a series of ten watches (all numbered) with gold bracelet. It is also a "complication" model with complete calendar and moon phases. 27P movement (1986).

Above: The mechanical movement of the 27P caliber. The very high quality of the manual finishing is evident, the angles polished one by one and the plates finished in the "côtes de Genève" design. In addition, the movement has five adjustments in order to ensure a constant degree of accuracy.

Below: A coin watch housed in an American twenty-dollar piece dating from 1893. The cover opens by pressing a button placed at the third hour. 9P mechanical movement (1982).

Above: The "Piaget d'or," a coin designed by Hans Erni to pay homage to Tradition, Harmony, and Peace. At La Côte-aux-Fées, the coin (struck in four versions: the half-, one-, two-, and five-"Piaget gold pieces") was transformed into a superb pocket watch with champagne dial and 4P mechanical movement (1984).

Below: Two hunters, the one on the left displaying a Chinese scene on its gold case, the other wrapped with four ribbons of pavé set-diamonds, 848 in all. Both watches have the mechanical 9P movement and were made in 1981.

Even Cary Grant, the great seducer of film, has been seduced in turn – by a Piaget watch whose special characteristics Yves G. Piaget is explaining.

Special Models and One-of-a-Kind Creations

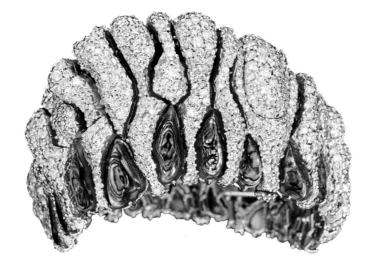

This bracelet watch, designed for Piaget by Jean-Claude Gueit, won the Diamonds International Award in 1969. It carries 1,113 diamonds and 21 mother-of-pearl "oysters." The watch is hidden in the central link. 6N1 movement.

Everything that emerges from the Piaget workshops with the company signature is exceptional, due as much to the quality of the materials used as to the fineness of execution. However, certain models are rare, and it would be hard to find their like among the entire world production of not only watches but also jewelry. These are the "special models" and the "one-of-a-kind creations": watches and parures often made on order for very special clients

The very unusual, unique pocket watch called "Snake charmer," in chased gold and enamel, was made in 1956-57 by Piaget in collaboration with Louis Cottier and Carlo Poluzzi. When the button under the pendant is pressed, the snake rears up to indicate the hour and a bird flies up to point out the minutes. The back shows a lion hunt.

(heads of state, well-known figures in the arts, culture, and industry) but sometimes made independently, simply to show that Piaget can "always do better than necessary" and that the word "impossible" does not exist for the master watchmakers and the artists who design, do metalwork, and work with precious stones at Piaget. The special models and one-of-a-kind creations can thus be defined as "dreams become reality": the concrete realization, precious and sparkling, of an idea and a design. Under the deft hands of the goldsmiths, engravers, enamelists, and jewelers, designs drawn on a piece of paper, after thousands of hours of work, become actual objects: in many cases, these are watches worth many thousands of dollars. Other objects amaze by the originality of their lines, their artistic richness, or their cultural references. In every case, it is the love, the art, and the skill that goes into the making of these "tellers of time" that counts. Piaget is not interested in competing to make the most expensive watches in the world; rather, it takes up the challenge, aimed at the entire watchmaking industry, to find the best way to combine richness and good taste, elegance and precision, originality and innovation. The most important pieces are finished at Prodor, but the entire experience acquired in the art of creating one-of-a-kind pieces is not confined to the walls of the luminous workshop in Geneva: it is applied to the development of new collections, to the effort to make a watch always thinner, for example, or always smoother to the touch even when it is entirely set with precious stones. This is why a Piaget is never an ordinary watch. Each model has the benefit

A coin watch made with a ten-dollar coin of 1894, with a one-of-a-kind movement only 1.35 mm thick (1955).

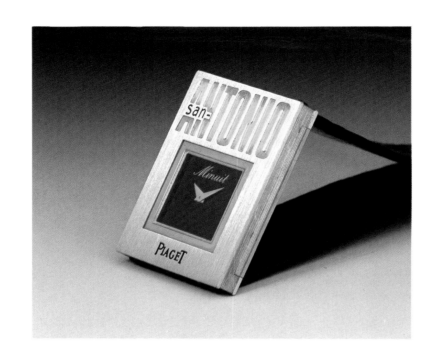

The "book watch" made in 1975 at the order of Frédéric Dard, author of the "San Antonio" detective novels. The case (which measures 26 by 35 mm, with a thickness of 4.9 mm) is in yellow, white, and pink gold, and the dial is onyx. The crown is on the left. 4P movement.

of exclusive technical and aesthetic solutions that make every piece "almost unique." For example, Piaget was one of the first companies to present coin watches, in which the timepiece has a case made out of a coin or a medal. This specialization, which goes back to the 1920s, provided a certain kind of manufacturing experience that led to numerous achievements in the area of the extra-thin. To take a coin and turn it into a watch is an extremely difficult technical undertaking: first, the obverse of the coin, which becomes the cover of the watch, must be cut off; then the coin must be hollowed out enough to fit the watch inside. Once these operations are carried out, the skill of the jeweler comes into play: the cover must be given invisible hinges so the coin does not betray

The Piaget "Phoebus," which at the beginning of the 1980s was the most expensive watch in the world, was made for a Japanese client. The case and the bracelet have 296 diamonds of the highest purity, for a total of 87.87 carats, set in platinum.

The bracelet watch "Hegira," made in 1980 in a limited and numbered edition for an Arab client. The medal covers an SP-caliber quartz movement. Both medal and bracelet exhibit an unusual asymmetry.

the transformation it has undergone, but it must open easily to reveal the time. To this end, the goldsmith fashions a tiny button in the milled edge of the coin. Almost invisible and perfectly aligned with the edge of the case, it releases a spring-loaded hinge once pressed, which unlatches the case. Such work requires great technical skill and calls for ultra-thin movements of very small diameter (exactly like those made by Piaget), since these tiny masterpieces, once completed, carry no sign of all this labor at first glance. So specialized is the skill required that even today, in this high-tech age, coin watches are mostly made by hand. The sole concession to modernity is the laser beam, which can cut the cover perfectly, without the slightest flaw, an operation that had been performed by mechanical instruments for many years. Aside from the goldsmithing skill they contain, certain one-of-a-kind pieces by Piaget have created a sensation in the auction houses by virtue of their great mechanical complexity. One such piece was the "Grande Sonnerie" model made about 1955. This bracelet watch contains a mechanical movement made early in the century (1910) that repeats the minutes. The minute repeater is actually one of the most difficult as well as fascinating technical "complications" in watchmaking. By pressing the button on the left side of the case, the watch strikes the hour indicated at that moment on the dial in one tonality, then in different tonalities strikes the quarter hour and the additional minutes that have elapsed. Two examples will clarify the operation: if the hands on the dial indicate

This skeleton pocket watch displays a tour-de-force of jewelry craftsmanship applied to watchmaking. The transparent case, made of gold and rock crystal, reveals the movement inside, which itself has been hollowed out and the remaining structure finely chased – an operation carried out by hand, and only by the most skilled engravers. The watch is further decorated with 282 diamonds (1981).

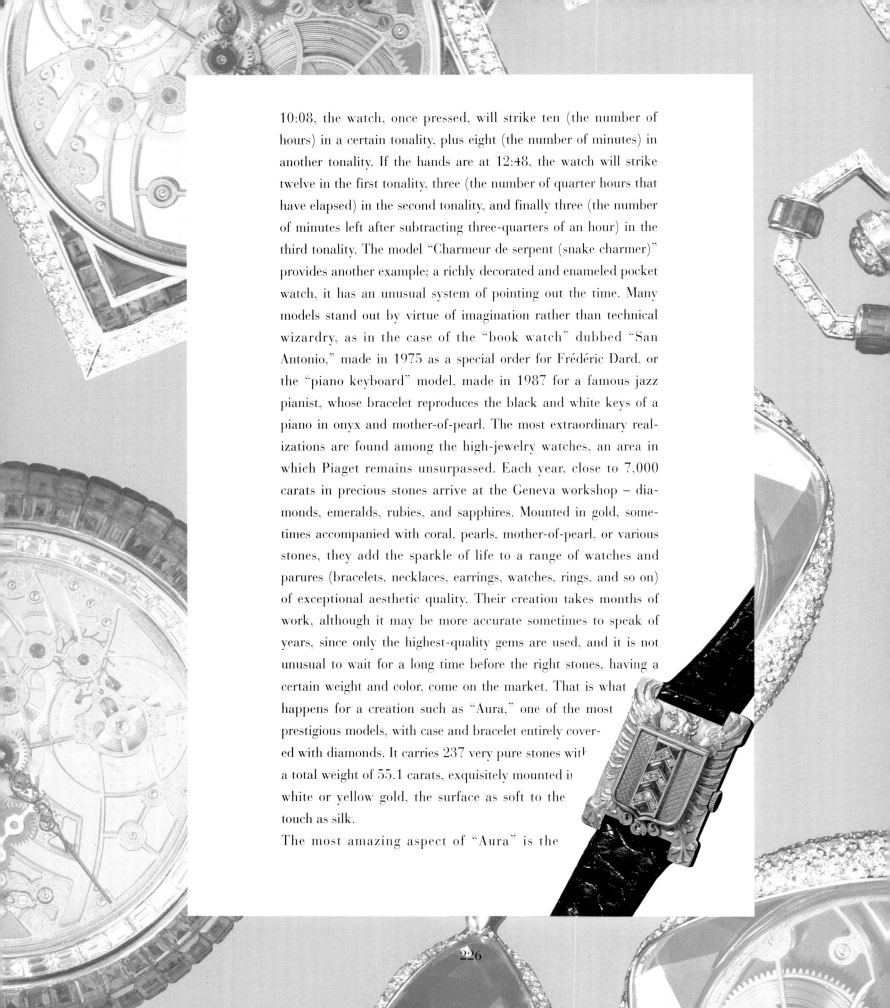

10:08, the watch, once pressed, will strike ten (the number of hours) in a certain tonality, plus eight (the number of minutes) in another tonality. If the hands are at 12:48, the watch will strike twelve in the first tonality, three (the number of quarter hours that have elapsed) in the second tonality, and finally three (the number of minutes left after subtracting three-quarters of an hour) in the third tonality. The model "Charmeur de serpent (snake charmer)" provides another example; a richly decorated and enameled pocket watch, it has an unusual system of pointing out the time. Many models stand out by virtue of imagination rather than technical wizardry, as in the case of the "book watch" dubbed "San Antonio," made in 1975 as a special order for Frédéric Dard, or the "piano keyboard" model, made in 1987 for a famous jazz pianist, whose bracelet reproduces the black and white keys of a piano in onyx and mother-of-pearl. The most extraordinary realizations are found among the high-jewelry watches, an area in which Piaget remains unsurpassed. Each year, close to 7,000 carats in precious stones arrive at the Geneva workshop – diamonds, emeralds, rubies, and sapphires. Mounted in gold, sometimes accompanied with coral, pearls, mother-of-pearl, or various stones, they add the sparkle of life to a range of watches and parures (bracelets, necklaces, earrings, watches, rings, and so on) of exceptional aesthetic quality. Their creation takes months of work, although it may be more accurate sometimes to speak of years, since only the highest-quality gems are used, and it is not unusual to wait for a long time before the right stones, having a certain weight and color, come on the market. That is what happens for a creation such as "Aura," one of the most prestigious models, with case and bracelet entirely covered with diamonds. It carries 237 very pure stones with a total weight of 55.1 carats, exquisitely mounted in white or yellow gold, the surface as soft to the touch as silk.

The most amazing aspect of "Aura" is the

On this page: The exceptional table clock "Carrousel," with rotating dial. The movement is placed in the base (around 1957).
Opposite: A one-of-a-kind piece of great beauty, the watch with concealed dial has a chased cover decorated with rubies and diamonds. It displays the coat of arms of the client's noble family. Mechanical movement (1950s).

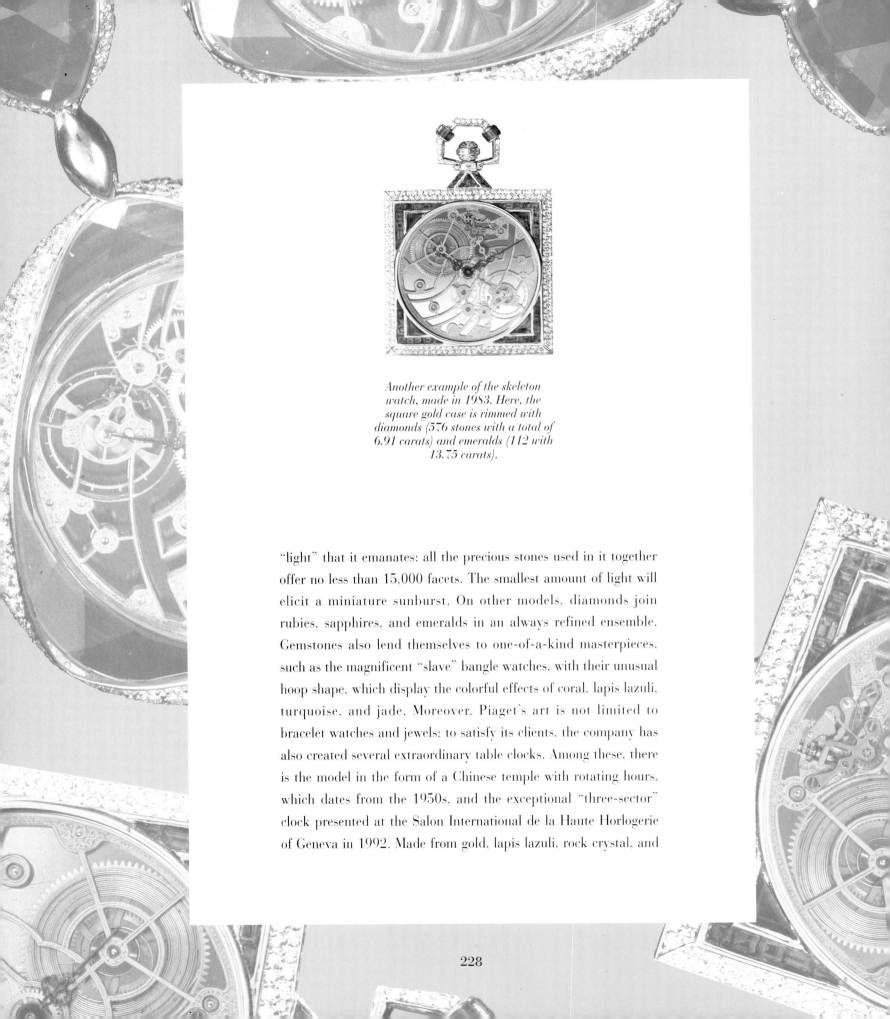

Another example of the skeleton watch, made in 1983. Here, the square gold case is rimmed with diamonds (576 stones with a total of 6.91 carats) and emeralds (112 with 13.75 carats).

"light" that it emanates: all the precious stones used in it together offer no less than 15,000 facets. The smallest amount of light will elicit a miniature sunburst. On other models, diamonds join rubies, sapphires, and emeralds in an always refined ensemble. Gemstones also lend themselves to one-of-a-kind masterpieces, such as the magnificent "slave" bangle watches, with their unusual hoop shape, which display the colorful effects of coral, lapis lazuli, turquoise, and jade. Moreover, Piaget's art is not limited to bracelet watches and jewels: to satisfy its clients, the company has also created several extraordinary table clocks. Among these, there is the model in the form of a Chinese temple with rotating hours, which dates from the 1950s, and the exceptional "three-sector" clock presented at the Salon International de la Haute Horlogerie of Geneva in 1992. Made from gold, lapis lazuli, rock crystal, and diamonds, it required 3,560 hours of work. It has three dials: on the two lateral dials, the sun alternates with the moon to indicate, respectively, the hours of the day or night, while on the central dial the time is marked by a telescopic hand in gold and diamonds that stretches out or shortens as it travels the arc around the dial.

The "keyboard" watch was made in 1987 for the well-known jazz musician Errol Garner. The bezel and bracelet, in mother-of-pearl and onyx, reproduce the alternating white and black keys of a piano. SP quartz movement.

This skeleton pocket watch is among the most amazing creations of Piaget. Benefiting from an extremely refined technique of "invisible setting" and a choice selection of stones, the circle of five rows of baguette-cut rubies that surround the case present a rounded and smooth surface. The watch has a total of 108 diamonds and 313 rubies (1985).

1988...

The drawing shows how Piaget interpreted the theme of the watch with hidden dial in the 1990s. The ribbed case is decorated with a cabochon. The cover is opened by means of a secret button. Quartz movement.

... A New Era Begins

On 26 April 1988, in the course of a well-attended press conference, Cartier announced the acquisition by its group of the companies Piaget and Baume & Mercier. An interesting aspect of the event was its simultaneous broadcast by satellite to seven important European cities – Paris, Brussels, Geneva, London, Madrid, Milan, and Munich – allowing invited journalists to address questions directly to Alain-Dominique Perrin, president of Cartier International. This audiovisual performance, rare indeed in the hermetic world of fine watchmaking, says much about the new approach of the group: to reach Europe and, from there, the rest of the world with "world-oriented" marketing and communication techniques, a strikingly innovative course in light of the business's traditional practices.

The official press release explained the reasons for and the methods of this operation: "Faced with increasing competition on the world market, the two groups, by pooling their potential techniques and production, their distribution capacities and their financial strength, consolidate their own positions and thus gain a clear advantage in the watchmaking sector. If the superior quality of Piaget's watchmaking has aroused Cartier's interest, the modern business and distribution methods of Cartier has intrigued Piaget. In uniting Cartier's century and a half of tradition with Piaget's exceptional creativity, the two great houses expect to carve out the lion's share of the world market in this sector." The merger proved extremely fruitful when after 1991 a recession followed the boom. Then, it took a strength and efficiency comparable to that of the new enlarged empire of Cartier to ride out a recession that felled more than one company and would sweep out many more before it ended.

The mystery clock with three separate dials was presented in Geneva in 1992. Its fabrication called for 3,560 hours of work. The gold and silver case is covered in lapis lazuli, and the minute hand is telescopic. 49 x 19.4 x 7.6 cm. Quartz movement.

Cartier did not take its motto, "The art of being unique," for granted. The top jeweler in the world at the time, it was also second in the production of luxury watches. Its international system of distribution covered 124 countries in 154 stores and 2,000 special dealers in watches, its principal markets being located in Europe and the United States.

With a production of about 20,000 watches a year, the Piaget Manufacture, founded in 1874, has counted among the world leaders in luxury watches since it introduced the famous ultra-thin 9P caliber in 1957 and decided to put its name only on watches in gold, platinum, or precious stones. Its goal since then has remained clear: to remain on the peak of the watchmaking pyramid by enriching the precious art of jewelry with the technical know-how and ability that had always given Piaget a place among the greatest names in Swiss watchmaking, such as Patek Philippe, Vacheron & Constantin, and Audemars Piguet. Relying on some hundreds of specialists established in strategic areas, its distribution, like its production, is the most selective in the world. Its elitism has been reinforced over the years: the Piaget store on the rue du Rhône in Geneva, which for thirty years served as a showcase of "those who count," has engendered a series of exclusive boutiques, first in Europe (Paris and Monte Carlo), then in Asia (Hong Kong, Singapore, and Kuala Lumpur).

At the basis of the agreement between the two houses was the meeting between two representatives of the international financial aristocracy: Yves G. Piaget and the Rupert family, the principal shareholder of the Richemont group. Besides Cartier, a beacon in French-European culture and history, this important group can boast among its holdings such prestigious names as Dunhill, Montblanc, Chloé, and Sulka. The

Above: The windows of the Piaget boutique in Monte Carlo are carefully lit to highlight the features of the watches and the precious stones. Right: The entrance to the Piaget boutique in Geneva. The elegant interior decoration is aimed at providing clients with the most comfortable of settings.

encounter between Yves G. Piaget and Johann Rupert blossomed with the realization of not only mutual interests and strategies but also philosophies of the vocation of true luxury, beyond fashions and trends: "A culture, a continuous search for the highest quality, the assurance of exclusivity, the love of noble materials: the whole crowned by a solid tradition of skills."

In taking responsibility for this alliance, the representative of the fourth-generation Piagets was already looking to the future: new techniques of marketing, communications, and distribution that only an empire like Richemont could guarantee to a company like his, a prestigious company whose production must remain exclusive in order to retain its superior quality.

In May 1988, the historic occasion was thus seized, "but not," Yves G. Piaget specifies, "because our sales had gone down or of any financial problems. The reasons for such a choice are very far from those that might have been suggested by some competitors. The Maison Piaget has always been its own banker. We have always reinvested, we have always underwritten ourselves." Fiercely "bankophobic," the Piaget family has always refused to borrow money, no matter what, thus guaranteeing its company an extremely sound business footing. This attitude explains the difficulty Yves G. Piaget had in convincing his family of the necessity to adopt an approach more in keeping with the international diffusion of the company. "It is important to inject new blood into our group; that is, only new people with a global vision, more international, will be able to assure our future well-being."

The agreement was based on two essential conditions. First of all, the company would keep its own identity and make its own decisions regarding the development of products, communication, and distribution, without, of course, excluding beneficial tactical

A brilliant demonstration of the talent of Piaget's artisans, this fine high-jewelry watch for men has an allure that reaches beyond the dial. Even the edge of the case is embellished with diamonds, which have a total weight of 22.3 carats. 4P mechanical movement.
Page 239: One of the celebrated "Aura" models. It is set entirely with diamonds, for a total of 28.6 carats, in an exceptional mounting.

combinations. Secondly, it must serve as the leading light ("the point of the diamond") of the prestigious names with which it became associated.

The generational conflict within Piaget resolved itself with the guarantee that the company would retain complete autonomy within the group and that the company's history and culture would be respected. With the advent of the new principal shareholder, Yves G. Piaget retained the position of president, a post he continues to hold, and has served as an ambassador or, as he puts it, "Minister of Foreign Affairs" for the company. Piaget's integration within the Cartier group has left it complete freedom in the area of creation, marketing strategy, and communication. Instead of continuing to favor the American and Far Eastern markets – which latter have returned to the fore with great vigor – it has turned its attention to creating a selection of styles capable of seducing the European market.

In light of the first products to emerge from this new era, it is safe to say that Piaget's creations remain faithful to the two complementary aspects of its tradition: on the one hand, perfectionism and recourse to high technology in the making of its movements and, on the other, creative imagination, the rare art of its jewelers and the continual search for the unconventional, an attitude that awarded Piaget in the 1960s with the reputation as the "jeweler in watchmaking." Such is the vocation of the watchmaker of La Côte-aux-Fées and the Geneva jeweler that is celebrated by the fabulous model "Aura": 152 diamonds (28 carats) for the ladies' version, 277 diamonds (61 carats) for the men's model, set in gold (case and bracelet) in an ingenious mounting.

In 1989, the end of a decade is at hand – that of the fax and the cellular telephone, of appearance, which has seemed to take precedence over essence... until the advent of neo-punk. In those strange years, sometimes described as "post-dada," excess in every genre and ostentatious displays appeared to mask the mysterious spasms that attend the changing of the millennium. The decade that succeeded it immediately assumed a less frivolous tone, diverted from the whims of changing fashions by the recession and the Gulf War. Plunged into gloominess, checked in its behavior, the world has sought to cushion its distress through a return to basic values.

Introduced in 1990 as a line of both watches and jewelry, the "Tanagra" collection stands as the symbol of Piaget's renewal. The extremely creative design of the watch stands out by virtue of two essential features: the curved ribs, which confer shape and character on a large case, and the flexible, articulated gold link, which allows the bracelet to adapt perfectly to the shape of the wrist. This feature, both structural and decorative, would become the distinctive mark of the collection, which also includes necklaces, bracelets, rings, and earrings.

Borrowed from a Greek village made famous in the fourth and third centuries B.C. for its elegant fired-clay figurines, the name Tanagra is significant in itself. By making the association with this ancient tradition of classical grace and lightness in a period marked by the utmost vulgarity in the world watchmaking production, Piaget intended to reaffirm a specific choice in the matter of taste and field of activity. It is not by chance that its publicity campaign for the previous year adopted as its slogan "flagrant discretion." Faithful to its vocation, true luxury thus takes, in advance of its time, the lasting paths of distinction and sobriety.

The sumptuous gala organized in honor of its premiere confirmed the importance accorded the new collection. For the occasion, a Greek temple of majestic proportions was reconstructed in the park of a magnificent château discreetly tucked away on the

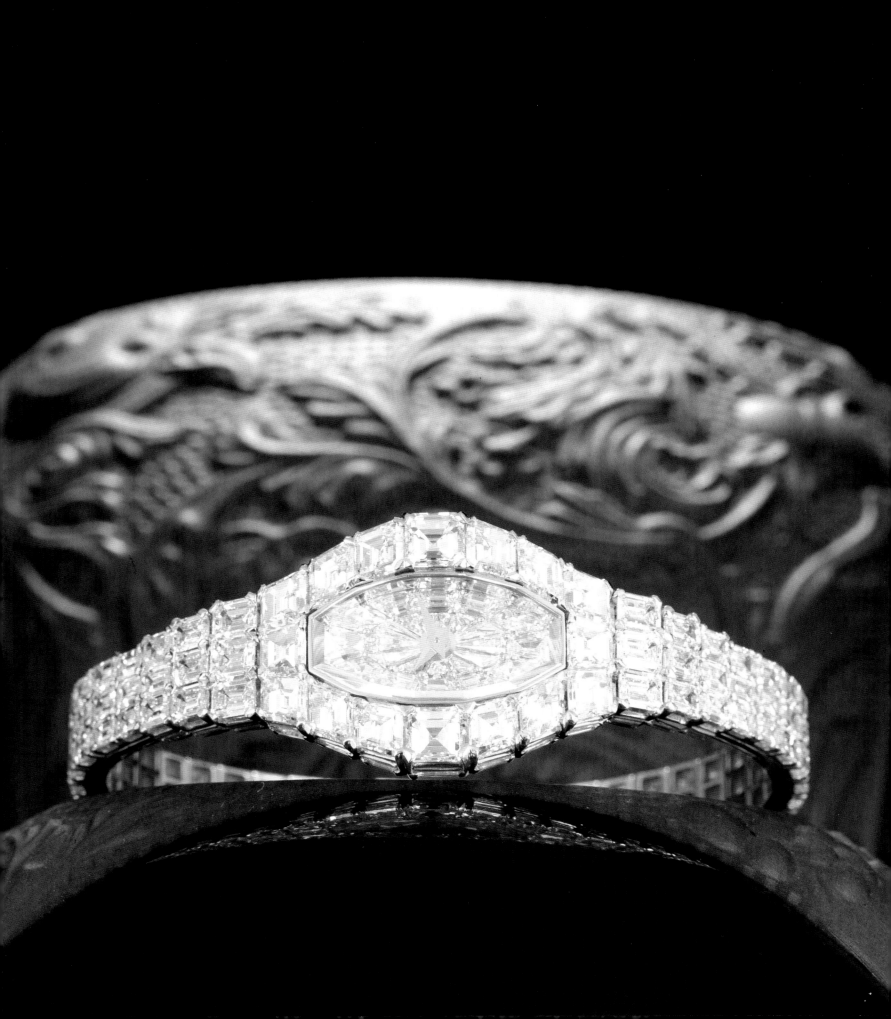

shore of Lake Léman. Inside, amid a decor of columns, friezes, and bas-reliefs, four life-size statues recalled the famous Tanagra. The spectacular fireworks that concluded the evening had nothing to do with the dying sparks of F. Scott Fitzgerald but rather offered dazzled eyes the reassurance that true luxury, indelibly inscribed in the human spirit, will never die.

While the model for men is truly impressive, the great novelty of "Tanagra" lay in the very special way in which Piaget approached the ladies' model, wedding aesthetic refinement to technical refinement. The jewel of the collection, undoubtedly, is the jeweled "Tanagra": a total of 524 diamonds (17.8 carats) decorate the men's version (which carries a mechanical movement) and 372 diamonds (21.4 carats) bedeck the ladies' model (with an electronic movement). The "Tanagra" line is also among the richest in variations: a model in satin-finished gold is available for both men and ladies; a square "Tanagra" (a great novelty at the Salon International de la Haute Horlogerie of Geneva in 1991); the round "Tanagra" with automatic movement; and, for those keen on the technical aspect of great watchmaking, the "Memorized Complication" and the "Tanagra Chronograph." With these two last models, which confirm the great watchmaking traditions in the realm of the jewel-watch, Piaget reaffirmed its aim to be the jeweler of la Grande Horlogerie.

In 1991, a revolutionary event turned the jealously

The rings and bracelet of the "Tanagra" collection echo the lines of the watch. The collection features interchangeable motifs decorated in diamonds and sapphires.

guarded world of fine watchmaking upside down. The united Cartier group privately proposed to some of the great names in the business that they display and sell their collections at Geneva, more elegant and above all better equipped from a logistical point of view than Basel, which holds the international watch and jewelry fair. The purpose was to separate the prestigious products from the rest of the watchmaking production. But it seemed equally essential that important international clients be well received in the neighborhood of the exhibition's site.

The idea was not new. Starting in 1983, Yves G. Piaget had made contact with several of the major fine watchmakers, designers, and hotel owners with the goal of founding in Geneva an international exposition of luxury products. For political reasons, the project did not come to fruition.

The proposal was simply to offer a preview of the best of the watchmaking production before going on to exhibit at Basel, where Piaget and Cartier retained separate displays the first year. When the invitation evoked no response from the other great names in watchmaking, Piaget and Cartier decided to go it alone, welcoming, besides Baume & Mercier, outside companies characterized by a very limited production, marked by the seal of an exceptional talent for watchmaking: Gérald Genta, Daniel Roth, and Frank Müller. The first SIHH (Salon International de la Haute Horlogerie) proved a great success in terms of both image and business. In the press conference he

Above: Blowup of a men's "Tanagra" model displaying the collection's characteristic curved ribs on the case and the bracelet. The quartz movement has a date function.

Right: The "Tanagra" chronograph, with fly-back function and perpetual calendar, carries the thinnest movement ever made with such "complications" – the 212P quartz movement.

Opposite: More jewels from the "Tanagra" collection with interchangeable cabochon motifs.

The international gala organized to introduce the "Tanagra" model. From left, Yves G. Piaget, Francesca Dellera, Ursula Andress, Marie Laforêt, Eddie Barclay, the actor Ben Gazzara, Yves Saint-Martin and Alain-Dominique Perrin, president of Cartier International.

held at the conclusion of this first experiment, an elated Alain-Dominique Perrin reported that the sales figures realized by the Cartier group during the first two days of the Salon exceeded those of the entire following week at Basel. The SIHH was able to take advantage of a wonderful setting: a space at Palexpo had been decorated in the form of a watch dial beneath a pale blue sky, as deliciously surreal as the backdrop of an American musical comedy of the 1950s. At the center of the space was a winter garden between Eastern-style white tents. Far from the crowds and activity of Basel, the professionals – to whom Piaget presented its "Dancer" chronograph and its round "Protocole" – found a setting conducive to the accomplishment of important business. It is clear to them that SIHH will become the showcase best adapted to the introduction of new collections and very special watches that forge the reputation and image of a great company.

The 1992 SIHH did not repudiate the success of the first Salon. On this occasion, the company presented the "Georges Piaget" watch, created in homage to its founder in a limited and numbered edition of 500, in yellow gold (numbers 1-200), white gold (1-150), platinum (1-100), and pink gold (1-50). Its very classic form recalled the traditional round shape of the 1940s and 1950s, the golden age of watchmaking favored by today's watch enthusiasts. The "Georges Piaget" watch borrowed its slightly domed dial and sapphire crystal, which gives it a special look, from

A "Tanagra" bracelet watch for ladies with a square case. The bezel is embellished with diamonds and the lapis lazuli dial has applied gold Roman numerals. The watch comes with either quartz or mechanical movement.

The Piaget pavilion at the Salon International de la Haute Horlogerie, held in Geneva since 1991.

Several rings from the "Possession" line. These are actually double rings, an outer movable ring that revolves around the inner ring.

the glorious 1940s. The piece is driven by the famous 9P-caliber mechanical movement with manual winding, which since its creation in 1956 had undergone numerous developments and served as the basic movement for all the watches of Piaget's classic line. Ultra-thin at two millimeters of thickness, it has a power reserve of thirty-eight hours. While its extreme thinness makes possible the use of greatly varied decorative techniques on its covering, the movement itself, richly engraved with plumes, scrolls, and roses around its rubies, contributes equally to the intrinsic value of the platinum series. All of the watches carry the signature of Georges Piaget on the back of their cases. Even the jewelry boxes that hold the watches, made of burled elm and lined with cream-colored leather, emphasize their beauty and value. All of this attention to detail no doubt explains why the watch, highly prized by collectors, can no longer be found.

What better way to commemorate the founder than to combine in a single watch the two essential aspects of the company's philosophy? Perfectionism ("Always do better than necessary") and the fierce intention to maintain the house's vocation of manufacture, that is, a place where, next to computers and the indispensable precision instruments, the human hand, sensitivity, and creativity continue to play a decisive role.

Also in that year, the "three-sector" table clock turned into a surprising extravagance that gave the company's watchmakers and jewelers a unique opportuni-

Above: A "Protocole" watch for men with engine-turned case (ultra-thin), and dial. Quartz movement.
Below: Three different versions of the "Protocole" model, with dials (from left to right) in onyx, lapis lazuli, and mother-of-pearl.

The "Georges Piaget," an extraordinary model created in honor of the company's founder, whose signature is engraved on the back of the case, was made in a limited and numbered edition (500 pieces: 100 in platinum, 50 in pink gold, 150 in white gold, and 200 in yellow gold). The movement is the famous mechanical 9P caliber, which – for the platinum version of this model – was entirely hand-chased.

ty to combine their talents. A central dial marks hours and minutes separately, while on either side two smaller dials display (at left) the course of the sun and (at right) that of the moon. From dawn to dusk and from nightfall to daybreak, the only clock of this type realized in these dimensions expresses the grandeur of Time through the most precious of materials: 18-karat gold, lapis lazuli, rock crystal, and, for the base, solid silver. Its exceptional nature, however, comes as much from its originality as from its commercial value (7 million francs).

In the autumn of 1992, the Maison Piaget offered to its admirers the rare opportunity to discover half a century of masterpieces from the company with the exhibition "Montres et Merveilles" (Watches and Wonders), based on the private collection that Yves G. Piaget and Emil Keller have continued to enrich with the acquisition of old models. After a successful debut in the prestigious space of the Palazzo Reale in Milan, it was shown at SIHH in 1993 before being welcomed at the Musée de l'Horlogerie in Geneva in 1994 and beginning a journey, starting in 1995, that will take it to important cultural locales around the world.

"Piaget, true values never change"... The current tendency to redefine basic values found an echo in the slogan adopted for the advertising campaigns of 1992-93. It signaled the triumph of essence over appearance, of the classic over the ephemeral, of the object bought for a lifetime over the impulse purchase, as well as greater attention paid to the true value of objects, to the way in which they are conceived and manufactured, to their historic legitimacy. The year 1993 saw the United States emerge from its recession, whereas Europe was still in the tunnel. The

On this page and opposite: The "Piaget Polo", another of the company's great successes. The case has small, fluted attachments and the bracelet features reeded central links. Automatic mechanical movement with date display for men and quartz movement for ladies.

Two moments from the "Courses d'Or," organized by Piaget, in Deauville (above) and Saint-Moritz (below) in 1994.

Asiatic tigers, including China, showed surprising signs of an economic boom. The luxury multinationals fell on them in the hope of acquiring small but significant pieces of the market by opening boutiques. This development had little effect on Piaget, which clearly had no need to retreat from such a challenge. The new developments in the "Gouverneur" and "Piaget Polo" collections were exhibited at the year's SIHH show, where the rectangular model "Belle Epoque" made its debut.

The idea of reinforcing connections between the great watch "couturier" and the equestrian world, which Yves G. Piaget had already forged with the Polo Championships held in the United States in the 1970s, was pursued in Europe. In the refinement of this world, Piaget saw a fitting symbol for its own production. Every year since 1989, the company was instrumental in organizing the now-famous Courses d'Or in Deauville, a resort town in France of enormous Belle Epoque charm, rich in memories. In winter, the action moves to the icy lake of Saint-Moritz in Switzerland, which welcomes the "White Turf" of he who, as president of the Association Genevoise des Compétitions Hippiques, has always experienced authentic admiration for the noble animal that is the horse. Crowned with success, these equestrian displays have been carried to Singapore, Hong Kong, and Milan, the beginning of an international career.

In 1994, Piaget celebrated the 120th anniversary of its founding at SIHH with a veritable festival of novelties, which, in the creativity of their design, the magnificence of their fittings, and their technical refinement, exhibit the full mastery of the grand master of watchmaking elegance. One of the highlights was undoubtedly the pocket watch with minute repeater and fly-back chronograph, a true masterpiece of watchmaking on the highest level covered in a brilliant coat of light: 60 diamonds on the bezel, 2 cabo-

Above: A "Gouverneur" for ladies in the version with diamond-set bezel.
Below: "A l'ancienne" model for men, with case in pink gold and dial with arched edges. Both watches have a mechanical movement.

A "Gouverneur" for men with date display and sweep seconds. It has a mechanical movement with automatic winding.

A chronograph version of the round "Protocole". White gold watch case with diamond-set bezel; blue subsidiary dials for the perpetual date display and the counters of this 212P quartz movement with fly-back function.

The "Gouverneur" in its chronograph version with three counters. Note the original attachment of the push-buttons to the case. Mechanical movement with manual winding.

On this page and opposite: An exceptional one-of-a-kind piece, created in 1994 to celebrate Piaget's 120 years and exhibited in Geneva. A chronograph pocket watch, it has flyback function and minute repeater. The watch is set entirely with diamonds. Its price is also unique: more than a million Swiss francs. Mechanical movement.

chon sapphires and 49 diamonds on the winding stem and crown, 76 diamonds on the shoulders, and 19 on the chain. Despite the flood of precious stones (264 diamonds with a total of 22.58 carats), the design, marked with the seal of the Piaget refinement, remains entirely classic. Another one-of-a-kind piece commemorated Piaget's anniversary: the bracelet watch with minute repeater, manual winding, and small seconds counter at the ninth hour, which glittered with the fire of 111 diamonds and 59 sapphires. More affordable in its price range, the commemorative medal watch with extra-thin mechanical movement and manual winding came in a white gold case and textured sunburst pattern on the dial with applied gold hours. Piaget's coat of arms appears on the obverse, while the reverse, finely engraved, displays the manufacture of La Côte-aux-Fées. Only 4 millimeters thick, the piece gave rise to a limited edition of 50. The latest additions to the "Tradition" collection are four special series, each consisting of 120 numbered pieces in yellow, pink, and white gold and platinum. Their special feature is that they carry the first Piaget mechanical movements with manual winding and power reserve display.

The watch "à secret," another feat both technical and aesthetic, is worth a special mention. Simply pressing on the mysterious lateral rib opens the cover to reveal a beautiful replica of Piaget's first watches of the 1940s and 1950s. This futurist line offers a men's model in gold with 18-karat-gold cabochon

Opposite: A group of designs representing several of the new watches made for Piaget's 120th anniversary and shown at the Salon International de la Haute Horlogerie in 1994. Clockwise from upper left: a commemorative medal in white gold hides watches with mechanical movements; chronograph with two counters, with cushion-shaped case and automatic mechanical movement; a one-of-a-kind piece with minute repeater; two of four models in a limited edition carrying a mechanical movement with manual winding and power reserve indicator.

On this page: The watch sketched on the page opposite at lower right became reality. An object not only beautiful to see but also to hear, this one-of-a-kind piece has a minute repeater and a case set with diamonds, while the bezel is emphasized by a circle of sapphires. Dial in gold, diamonds, mother-of-pearl, and sapphires. Small seconds counter at the ninth hour.

and a ladies' model with cabochon sapphire or emerald. Leaving no corner of fine watchmaking unexplored, Piaget also presented its skeleton chronograph with hand-chased finish – obviously, in a limited edition. More classical in conception is the mechanical chronograph with two counters, endowed with a shaped case and available in yellow or pink gold. This catalogue of the models shown at the SIHH cannot be concluded without mentioning the piece that is so characteristic of the Piaget style, it reveals itself at a glance: the "rectangle à l'ancienne" in the version decorated with diamonds, with the center of the dial in onyx and its outside with diamonds.

In commercial terms, the SIHH enjoyed a great success: seven thousand clients and visitors, four hundred journalists from around the world, and professionals from over sixty countries.

Anticipated revenues proved equally rewarding, thus confirming that the recession does not affect true luxury products and that there is always room for true values: beautiful craftsmanship, realized with passion and ability; the traditional arts of old Europe, including jewelry and watchmaking; new expressions in design and technological progress ... always placing people at the heart of the enterprise, as demonstrated by the history of this small "great family" of peasants who, since their beginnings in the remote mountains of the Swiss Jura, have conquered great personalities, markets, and the heights of international elegance.

On this page: The 35-mm model "Tradition" displays very classic lines and a round case. Quartz or mechanical movement.

Opposite: Three models with "square curved" case; it comes with white dial or dials made of gemstones.

Page 262: Another model of fine watchmaking, a skeleton chronograph with oval case decorated with diamonds or, on request, rubies and sapphires.

PHOTOGRAPH CREDITS

Mondadori Press/Angelo Cozzi
Jean-Luc Brutsch, Geneva
W. Abplanalp, Neuchâtel
Publi Conseil SA
Photos RKM
Photos NBC SA
Roger-Viollet
Gamma-Liaison/Michael Abramson
Art Photo/Denis Hayoun
Thierry Bourdeille
J. F. Schlemmer
Henri Rossier
Maurice Aeschimann
Zefa Italiana/Hans Blohm
Grazia Neri/Rapho
Charles Guyot
Yves Coatsalion

Archives Historiques Piaget
Archives de l'Etat de Neuchâtel
Archives de la Société des Nations, Bibliothèque des Nations Unies, Geneva
City of Geneva, Documentation photographique, Collection W. Aeschlimann
Musée des Beaux Arts, La Chaux-de-Fonds
Antiquorum, Geneva
La Côte-aux-Fées by E. Quartier-La-Tente, Editions J. J. Kissling, Neuchâtel
Le Canton de Neuchâtel by E. Quartier-La-Tente, Editions Attinger Frères, Neuchâtel

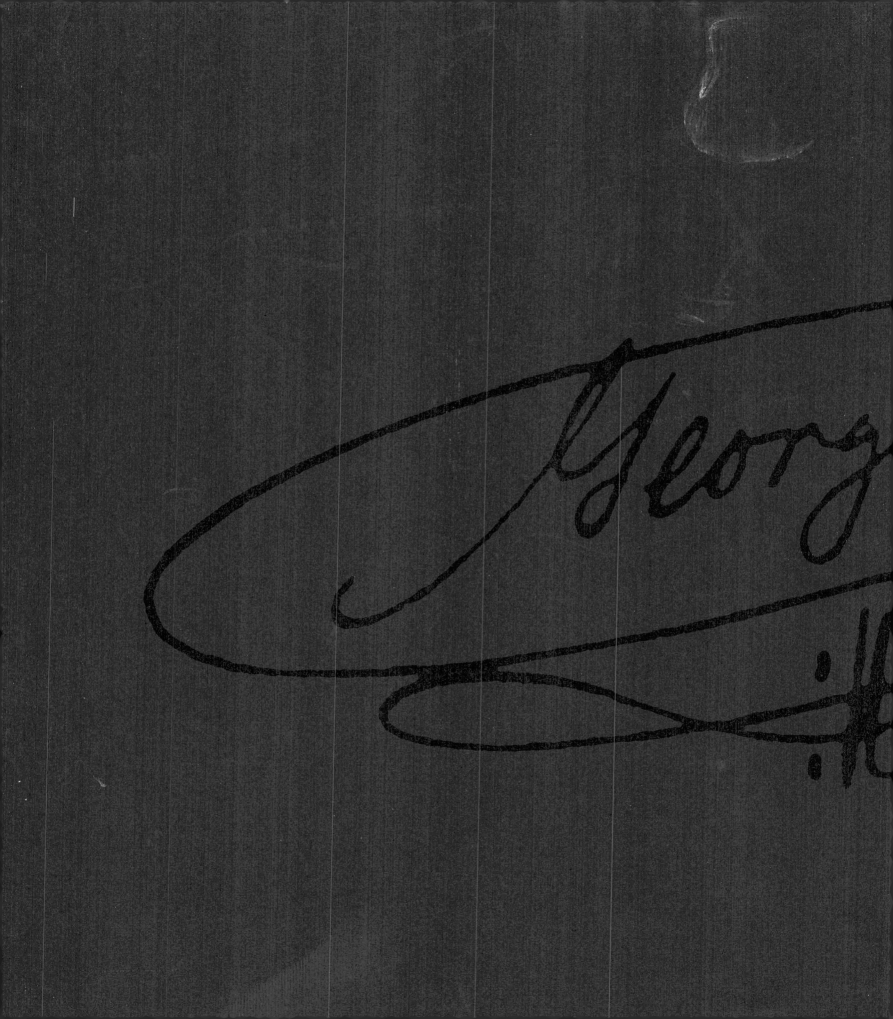